PETROC™

LEARNING TECHNOLOGIES & LIBRARY SERVICES

Please return this item on or before the last date shown below

2 1 OCT 2014

North Devon Campus: 01271 338170
Mid Devon Campus: 01884 235234
Email: library@petroc.ac.uk

General Preface to the Series

Because it is no longer possible for one textbook to cover the whole field of biology while remaining sufficiently up to date, the Institute of Biology proposed this series so that teachers and students can learn about significant developments. The enthusiastic acceptance of 'Studies in Biology' shows that the books are providing authoritative views of biological topics.

The features of the series include the attention given to methods, the selected list of books for further reading and, wherever possible, suggestions for practical work.

Readers' comments will be welcomed by the Institute.

1984   Institute of Biology
                                              20 Queensberry Place
                                              London SW7 2DZ

Preface

The rapid progress in molecular genetics in recent years has been made possible because geneticists have chosen to work with very simple organisms. Research had been concentrated on bacteria and their viruses because they are technically much easier to handle than higher animals or plants. However there are fundamental differences between the molecular biology of unicellular prokaryotes and that of multicellular eukaryotes, and so we obtain an incomplete picture by concentrating on the former. The question therefore arises: are there ways of applying bacterial genetics techniques to higher organisms? This book explores the possiblity of using animal and plant cell cultures as 'honorary microorganisms' for genetical experiments. Several ways of producing new combinations of genetic material in them are explored. These studies have led on to the possibility of introducing new genetic material into whole animals and plants. Developments in these areas are not only providing knowledge of the genetics of higher organisms but also have important medical and agricultural implications.

I am grateful to Dr. John Sparrow for constructive criticisms of the manuscript of this book.

York 1984   J.R.W.

Contents

1 Introduction

1.1 The trend to simple experimental systems in genetics

Any student of the history of genetics is soon struck by the tendency of geneticists to work with progressively simpler organisms. Early geneticists were primarily interested in humans and their domesticated plants and animals, but by 1920 it was appreciated that *Drosophila* offered valuable technical advantages with its short life cycle and capacity to produce large numbers of offspring in a well defined environment. By the 1950s many geneticists were beginning to work with bacteria and their viruses for the same basic reasons. The trend to simpler, technically more convenient systems is now continuing with the development of recombinant DNA technology, where much of the work involves manipulation of individual classes of DNA molecules.

This tendency to choose simple experimental systems to solve difficult problems has been of great value to geneticists. Without any doubt, recent developments in molecular genetics could not have occurred if simple experimental systems had not been available. However, there is a disadvantage in working with these simple systems – namely that attention is no longer being directed to the organisms in which society as a whole is particularly interested.

Many people would actually prefer to study the molecular genetics of humans, higher animals and higher plants rather than the molecular genetics of the bacterium *Escherichia coli*. The reasons for this preference can be broadly subdivided into 'applied' and 'pure'. The applied reasons are more obvious and easier to explain. Firstly, knowledge of the molecular genetics of higher animals and plants may have valuable implications for agriculture. With 30% of the world's population undernourished, discoveries of ways of increasing food production must be of the utmost concern to the scientific community. Is it possible to manipulate the genetic material of domesticated animals or crop plants in new ways which would increase their yield or their resistance to disease? Is it possible that the nutritional quality of crop plants may be improved by genetic modification, for example making changes which would increase the levels of important amino acids, thus raising the quality of plant proteins?

Secondly, a deeper understanding of the molecular genetics of humans may contribute to the treatment of some forms of human disease. Many human diseases have a genetical component and rational discussion of their therapy can only be based on a sound understanding of their causes.

There are also important scientific reasons why attention should be directed to the molecular genetics of higher organisms. One of the most challenging problems in modern biology is to understand how gene expression is controlled during development. Understanding of the control of gene expression in bacteria does provide a valuable starting point for work on higher organisms, but it has become increasingly apparent during the last decade that there are fundamental differences between the control of gene expression in prokaryotes and eukaryotes. In a broader sense, development is largely confined to multicellular organisms and can only be realistically studied using them.

Although biological research is motivated by inquisitiveness about all living things, and the molecular genetics of bacteria is one of the most fascinating areas of biology, there is a natural human tendency to be particularly interested in our own and related species.

How could we learn about the molecular genetics of humans and their domesticated plants and animals in equivalent depth to bacteria? It may be possible, at least partially, to resolve this question by using cells of higher organisms, grown in culture, for genetical experimentation and manipulation. The genetics of cultured cells is developing rapidly and will continue to do so in the foreseeable future. The aim of this book is to present a snapshot of recent developments in several of the areas which involve the common basic approach of attempting to apply microbial genetics techniques to the genetics of higher organisms.

Some background information may be useful and the two remaining sections of this chapter are intended as a prologue for the following chapters of the book.

1.2   Microbial genetics techniques

Which microbial genetics techniques might be applicable to plant and animal cells grown in culture? One obvious point is that cell cultures do not have any kind of sex life and it is therefore not going to be possible to use the conventional sexual cycle of gamete fusion alternating with meiosis as the basis for genetical analysis. In order to bring about new combinations of genetic material for study it will be necessary to have some kind of alternative to sex.

Bacterial geneticists faced similar problems about thirty years ago. The bacterial genetic material is not segregated in rod shaped chromosomes during any form of meiosis, and there is no fusion of 'haploid' gametes to give rise to 'diploid' progeny. However, there are several important ways of bringing about exchange of genetic material in bacteria.

The first of these is *conjugation*. In conjugation, some genetic material passes from a donor bacterium to a recipient bacterium as a linear DNA molecule. This may therefore be regarded as a sexual process, although it does not involve the fusion of equivalent gametes. It appears to have no counterpart in animal and plant cells grown in culture and will not be considered further in this book.

However, two other long established methods of bringing about exchange of genetic material in bacteria may have equivalents in cell culture. These two bacterial techniques are called transformation and transduction (Table 1–1). During *transformation*, DNA is liberated by lysis of donor cells and it taken up by recipient cells. Small fragments of DNA may be incorporated into the

Table 1–1 Methods of bringing about genetic exchange in microorganisms which may be applied to animal or plant cells in culture.

BACTERIA	
Transformation	Free DNA is transferred from donor cells to recipient cells
Transduction	DNA is transferred from donor cells to recipient cells by viruses (bacteriophages)
Gene cloning	DNA is inserted in a vector, which is then inserted into bacterial cells. The vector is a small DNA molecule capable of being replicated.
FUNGI	
Parasexual cycle	Fusion of hyphae, followed by fusion of nuclei and then chromosome loss during mitotic divisions.

recipient's DNA, thus altering its genotype and phenotype. Cell-to-cell contact is not required. During *transduction*, DNA is carried from a donor bacterium to a recipient bacterium within the protein coat of a bacterial virus (usually called a bacteriophage or simply 'a phage'). The result is again a transfer of fragment of DNA from a donor bacterium into the DNA of a recipient with a resulting change in the latter's genotype.

The fourth (and most widely publicized) method for producing new combinations of genetic material in bacterial cells is *gene cloning*. Lengths of DNA (which may be from unrelated species) are inserted into DNA molecules (vectors) which have the capability to replicate inside bacteria. The vectors may be extrachromosomal circles of DNA called plasmids (see DAY, 1982) or viral DNA. Insertion of new DNA (i.e. the genes which are to be cloned) into the vector is done in the laboratory using biochemical techniques for cutting and rejoining DNA molecules which are often collectively referred to as 'recombinant DNA technology'. Because the vector replicates itself and its inserted DNA, large numbers of copies of the DNA insert (and, sometimes, large amounts of the protein coded for by the DNA) can be obtained by these methods.

The final mechanism for achieving genetic exchange in microorganisms which should be considered does not involve bacteria but occurs in certain fungi, such as *Aspergillus nidulans*. The overall sequence of events is called the parasexual cycle (Fig. 1–1) and may commence if two genetically different strains of the same species grow together. Occasionally hyphae fuse, giving rise to a heterokaryon, where two different types of nuclei are present in a common cytoplasm. Within this heterokaryon nuclei may fuse to double the chromosome complement. Techniques are available for isolating cultures of the fungus where this has happened. *Aspergillus* nuclei are normally haploid and so the new strain arising by nuclear fusion is diploid. The chromosome complement of such diploid stains of *Aspergillus* is not stable and during mitosis (note, not meiosis) some chromosomes do not move to the poles and are lost until a haploid chromosome number is attained once more. The resulting 'progeny' strains of *Aspergillus* contain the same *total* amount of genetic material as the two original 'parents', but in *new combinations*. Analysis of such 'progeny' of this 'parasexual cross' enables information to be obtained about the arrangement of genes on chromosomes.

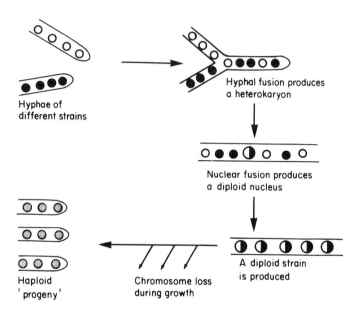

Fig. 1–1   The parasexual cycle in *Aspergillus* (see text for details).

1.3   The use of cell cultures in genetics and molecular biology

The first step in applying the techniques of microbial genetics to higher organisms must be to find ways of manipulating their cells in a similar fashion to those used for microorganisms. It is fortunate that techniques for animal and plant cell culture have existed for some time and much of the work described in this book involves applying genetical techniques to cells grown in culture. Techniques of animal and plant cell culture have been reviewed in SHARP (1977) and BUTCHER and INGRAM (1976). Some basic features of animal cell cultures will be mentioned in the following paragraphs and some basic features of plant cell cultures will be outlined at the beginning of Chapter 7.

It is important to appreciate the difference between 'primary' and 'established' animal cell cultures. When cells are first isolated from human or other mammalian tissues into culture medium, they normally undergo about 20 divisions before ceasing to grow. These initial cultures with a finite life span are known as *primary cell cultures*. Occasionally a cell line arises from a primary cell culture which has unlimited lifespan and which can be subcultured indefinitely like a microorganism. This is known as an *established cell culture* or *cell line* and is the type of culture used for most of the work described in Chapters 2, 4 and 5. (The nature of the underlying change between a primary and an established cell culture is still not clear; it may have some features in common with some of the early stages of development of cancerous cells.)

The general similarities and differences between working with bacteria and established animal cell cultures are also worth considering. The most obvious difference between the two when cultured in petri dishes is that bacteria are grown on the surface of medium solidified with agar, but animal cells grow attached to the glass or plastic surface, bathed in liquid medium. The growth rate of animal cells is much slower (doubling times of 10 to 20 hours compared with 20 to 60 minutes for bacteria) and their culture medium must be much richer (mainly due to the presence of added serum, which contains ill-defined, but necessary, growth factors). However, in many ways, established cell cultures can be manipulated and plated out as if they were bacteria (Fig. 1–2).

When geneticists realised that mammalian cells could be handled like bacteria, they began to investigate possible ways of exchanging genetic material between cell lines in some form of a cross. As described for bacteria in section 1.1 this need not simply involve the fusion of two haploid gametes. It turns out that there are several completely different ways of producing new arrangements of genetic material of cells grown in culture. Some of the main techniques are very briefly outlined here.

(*a*) *Cell fusion*   Mammalian cells in culture occasionally fuse and undergo a series of events, rather similar to the *Aspergillus* parasexual cycle (section 1.2). After cell fusion, the nuclei may fuse to give approximately double the number of chromosomes which were present in the original cells. If the cells which fuse are different cell lines, a hybrid is produced. During subsequent mitotic divisions, some chromosomes are lost, giving rise to progeny clones with different chromosome compositions. The study of these progeny clones can allow a form of genetic analysis in culture which is now providing a wealth of information about the arrangement of human genes on chromosomes (Chapter 2). Cell fusions involving certain kinds of antibody producing cells are of special interest for quite a different reason. They give rise to hybrid cell lines which produce very pure antibodies in large quantities. This discovery has important medical implications (Chapter 3).

(*b*) *Transformation*   DNA can be introduced into animal cells in culture either in the form of pure DNA or as isolated chromosomes, thus allowing a form of genetical study equivalent to bacterial transformation (section 1.2). Transformation is now proving useful in studying several fundamental biological problems. For example, it is being used for the characterization of human cancer genes in what is widely regarded as the most significant development in cancer research for many years (Chapter 4).

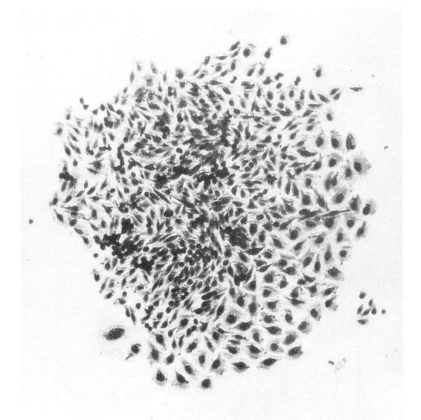

Fig. 1–2 A colony of Chinese hamster cells, grown in a plastic petri dish. The colony has arisen from a single cell after about nine rounds of division. (Courtesy of Mr M. Anderson, Biology Department, University of York.)

(*c*) *Gene cloning* Genes can be inserted biochemically into the DNA of cloning vectors which can replicate in animal cells. This enables gene cloning to be performed in animal cells in a similar way to bacterial gene cloning, and opens up new scientific and practical horizons (Chapter 5).

The techniques just mentioned involve producing new genetic complements in cells which are being grown in culture. Recently, however, there has been increasing interest in the possibility of incorporating new genetic material into whole animals or possibly even humans. This would have important consequences for animal breeding and may conceivably lead to gene therapy for human genetic diseases (Chapter 6).

Plant cell cultures can be manipulated in several ways which are broadly similar to, but have important differences from, the techniques used for animal cell genetics. A most important difference is that whole plants can be regenerated from cells which have been manipulated in culture. This could have extremely valuable implications for plant breeders and eventually for world agriculture (Chapter 7).

2   Hybridization of Mammalian Cells

2.1   Mutants in established cell culture

In order to begin the genetic analysis of mammalian cells growing in culture, it is necessary to have some mutants to study. Mendel would not have got far with the study of genetics of pea height if all peas were equally tall; it was only when he crossed tall and short peas that he began to get interesting information concerning patterns of inheritance of genes for height. For the same reason, the discussion of genetic manipulation of mammalian cells requires the consideration of the nature and origin of mutants available in cell cultures.

Many classes of mutants used in mammalian cell genetics are similar to those found in bacteria. In both cases, probably the easiest mutants to isolate are those which have increased resistance to a particular drug (Fig. 2–1). When normal 'wild type' cells are plated in increasing concentrations of a drug, a survival curve is obtained. Spontaneous mutants, which are able to grow on concentrations of the drug which would kill wild type cells, may arise within the cell culture (with a frequency of around 10^{-6}, although this frequency may be increased after the cells have been treated with chemical mutagens). These resistant mutants can be isolated from the general population simply by plating large numbers of cells (say 10^7) into a suitable drug concentration (concentration A in Fig. 2–1).

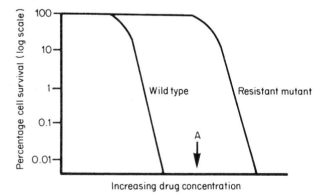

Fig. 2–1   Survival curves of wild type (*left*) and a resistant mutant (*right*) plated in increasing drug concentrations. Concentration A is a drug concentration which may be used for selection of resistant mutants.

The specific example of azaguanine resistance has been widely studied in cell cultures and, as shall be discussed later, it has several features which make it particularly useful in mammalian cell genetics. Azaguanine is an analogue of the purine, guanine (Fig. 2–2). If azaguanine is present in the growth medium,

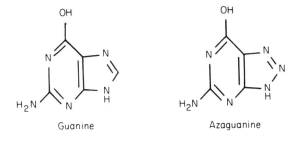

Guanine   Azaguanine

Fig. 2–2   The formula of guanine, a normal constituent of DNA, and its analogue, azaguanine.

cells incorporate it into their DNA and, as a consequence, are killed. One of the enzymes involved in this sequence of events is called hypoxanthine guanine phosphoribosyl transferase or HGPRT for short. This enzyme normally converts purines (such as guanine) into nucleotides (such as guanosine monophosphate), which can subsequently be used for DNA synthesis. In azaguanine resistant cells however, this enzyme has an altered structure, which prevents it from carrying out its normal function and consequently azaguanine is not incorporated into the cell's DNA. The mutant cell is therefore able to survive in the presence of azaguanine. The lack of HGPRT is not lethal to the cell because it has two pathways for nucleotide biosynthesis, the 'scavenger' pathway, described above, and an alternative pathway which involves their 'de novo' synthesis from sugars and amino acids (Fig. 2–3). (Azaguanine cannot be incorporated into DNA by the second pathway.)

Interestingly the same biochemical defect is observed in a human genetic disease called the Lesch-Nyhan syndrome. Individuals suffering from this disease lack HGPRT and have increased levels of de novo purine synthesis. This abnormal metabolism leads to clinical symptoms of spasticity, developmental retardation, uncontrollable movements of the limbs and self mutilation (by biting lips and fingers).

Although resistant mutants are the easiest to isolate in cell cultures, there are a variety of other types which are also useful. Mutants which have extra nutritional requirements are an important class. For example, although hamster cells in culture do not normally require the amino acid proline to be present in their growth medium, mutants may be isolated which do have such a requirement, because they are deficient in an enzyme involved in proline biosynthesis. Other nutritional mutants may be isolated which have requirements for other nutrients not normally needed by cells grown in culture.

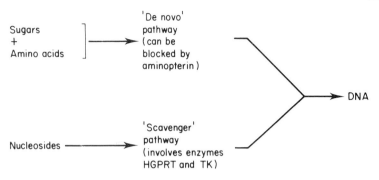

Fig. 2–3 Alternative pathways for nucleotide synthesis in mammalian cells. Azaguanine resistant mutants are defective in the enzyme HGPRT and therefore do not incorporate it into their DNA by the scavenger pathway.

One general feature of mutation in mammalian cell culture puzzled geneticists for some time. In coventional genetics of higher organisms mutations are thought of as being recessive or dominant. Most commonly, the mutant copy of the gene (i.e. the mutant allele) is recessive to the wild type copy (i.e. the wild type allele). It will become apparent in this book that when mammalian cells are fused (section 2.2) or when gene transfer is carried out in cell cultures (sections 4.2 and 4.3) the mutant alleles are usually recessive, as might be expected. However, there appears to be a paradox, because when the mutations are originally isolated in cell culture, these recessive mutations are isolated in apparently diploid cells. Why are they not masked by dominant wild type alleles, which might be expected to be present on the homologous chromosomes? It now appears that some mutations (including azaguanine resistance) are on the X chromosome, only one copy of which is present in male cells and only one copy of which is functional in female cells (the other copy being in a condensed, inactive form called the Barr body). Hence the problem of dominance does not arise. In some other cases, the explanation is probably that cells in established cell culture have usually lost some small regions of some chromosomes and if a mutation arises in an equivalent region of the intact homologous chromosome, the question of dominance again will not arise because there is only one copy of the gene present. Other more complex explanations have also been proposed.

2.2   Cell fusion and chromosome loss in hybrids

The first form of genetic analysis in cell cultures to be considered here is outlined diagrammatically in Fig. 2–4 and is sometimes called the parasexual cycle. It is rather similar in outline to the parasexual cycle in *Aspergillus* (section 1.2), but there are some important differences in the pattern of chromosome loss from the hybrid nucleus.

When two different cell lines are grown together in the same culture vessel,

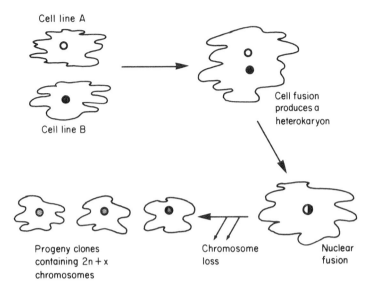

Fig. 2–4 The parasexual cycle in mammalian cell cultures. For explanation, see text.

cell fusion (Fig. 2–5) and nuclear fusion may occur to produce a cell line with a combined chromosome complement. Fusion can occur between cells of the same or different species. If the original cell fusion is between a primary cell culture of one species and an established cell line of another, there is rapid chromsome loss from the hybrid during subsequent mitotic divisions until clones stabilize with a chromosome complement somewhere above the diploid level (2n + x, where 2n is the diploid number of chromosomes and x is a small number). The details of the process and its use in human genetics are discussed below.

Multinucleate cells have been seen in vertebrates for over a hundred years. They are observed particularly frequently during infection with viruses such as human tonsils infection during measles. Multinucleate cells were seen in some of the earliest tissue cultures and by 1954 it had been shown that measles, mumps or para-influenza viruses cause cell fusion in tissue culture. In 1960, G. Barski demonstrated that mixed cultures of two cell types contained hybrid cells where nuclei of the two kinds had fused to give a single nucleus containing the chromosomes of both cell types. Soon after this, B. Ephrussi grew pure hybrid cell lines in culture. In 1965, H. Harris and J.F. Watkins isolated hybrids formed between cells derived from two different species (e.g. mouse and man). This observation has been of particular importance in human genetics, as will be discussed in the following pages.

Only a small minority of cells in a mixture of two types fuse together and hence techniques have been devised (*a*) to increase the number of cell fusions and (*b*) to select out the rare hybrids from the parental mixtures. The

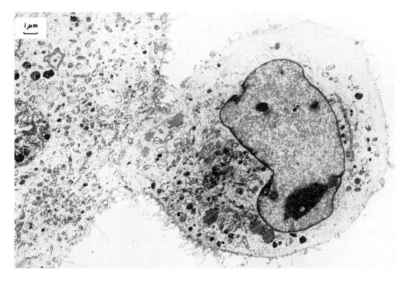

Fig. 2–5 The boundary between two mammalian cells which are in the process of fusing with each other. (Courtesy of Professor H. Harris, Department of Pathology, Oxford University, and Oxford University Press.)

commonest way to increase the frequency of cell fusions is to expose the cell mixture to ultraviolet(UV)-inactivated Sendai virus (a member of the para-influenza group of viruses). The virus appears to embed itself in the membranes of adjacent cells favouring their dissolution and consequently the formation of cytoplasmic bridges between neighbouring cells. Cell fusion can also be favoured by brief exposure to polyethylene glycol.

Selection of rare hybrids from the large excess of unfused cells can be achieved by devising growth media which will allow only the hybrids to survive. One of the earliest and most widely used of these is the 'HAT selection' technique devised by J.W. Littlefield. In this selection one of the parental cell lines is azaguanine resistant and hence lacks the enzyme HGPRT, (see p. 9). The other parental cell line, designated TK⁻, lacks a different enzyme in the scavenger pathway for nucleotide synthesis, namely thymidine kinase. Consequently neither of the two strains are able to utilize DNA precursors (e.g. hypoxanthine or thymidine) in the medium for DNA synthesis via the scavenger pathway. A drug called aminopterin is added to the medium which is an inhibitor of the de novo pathway (see Fig. 2–3) and the net effect of mutations *and* drug is that neither parental cell line can grow in HAT medium (medium containing hypoxanthine, aminopterin and thymidine). If cell fusion occurs, each parental cell type contributes a wild type copy of the gene which is defective in the other to the resulting hybrid. Thus the mutant which is TK⁻ has a functional gene for HGPRT and the other mutant which is HGPRT⁻ has a functional gene for TK. The hybrid is consequently heterozygous for the

HGPRT gene (i.e. it is HGPRT⁺/HGPRT⁻), and is also heterozygous for the TK gene (i.e. it is TK⁻/TK⁺). Since the mutant genes are recessive, the hybrid has a wild type phenotype, i.e. a functional scavenger pathway. Hybrid cells can therefore grow in HAT medium and be isolated from the parental cells, which do not grow.

Hybrids formed between cells of the same species or very closely related species are usually fairly stable. They divide vegetatively, with only occasional loss of chromosomes. However, in 1967 Mary Weiss made the surprising and subsequently very important observation that human–mouse cell hybrids lost chromosomes very rapidly. Furthermore, the human chromosomes were lost *preferentially*, so that after about 20 cell generations the hybrids contained all the mouse chromosomes and only 2 to 15 human chromosomes. Preferential chromosome loss continues until, after 100 to 150 cell generations, stabilization occurs at around 1 to 3 human chromosomes. (A similar preferential loss of chromosomes from one species occurs in other interspecific hybrids.) Different cell lines (or 'clones') arising by this sequence of events contain different human chromosomes. It should be clearly understood that these 'progeny clones' have arisen by chromosome loss during mitotic divisions and not by any kind of meiosis.

The net result of this sequence of events is to produce a variety of different combinations of human genetic material in the progeny clones. By studying these progeny clones geneticists have been able to take an entirely fresh approach to human genetics.

2.3   Mapping human genes

The first type of information which has been obtained using the parasexual cycle is the assignment of particular genes to particular chromosomes. The first human gene to be mapped in this way was the gene which specified the structure of the enzyme thymidine kinase (see p. 12). Human cells containing a functional gene for thymidine kinase (designated TK⁺) were fused with mouse cells with a mutation in the TK gene so that they lacked this enzyme (TK⁻). The hybrid (TK⁺/TK⁻) had functional thymidine kinase, showing that TK⁺ was dominant over TK⁻. When this hybrid starts to lose human chromosomes, all the resulting clones will have the mouse TK⁻ gene, but only some will have the human chromosome carrying the human TK⁺ gene and others will not. The clones which do have the human chromosome carrying the TK⁺ gene will have the genotype TK⁻/TK⁺ which gives the phenotype TK⁺ (since TK⁺ is dominant over TK⁻). The clones which do not have the human chromosome carrying the TK⁺ gene will have the genotype TK⁻ which gives, of course, a TK⁻ phenotype. (Fortunately, the TK⁺ and TK⁻ phenotypes can be distinguished in practice because TK⁻ also leads to resistance to the drug bromouracil.)

Progeny clones can then be examined and classified on the basis of two different criteria. Firstly, they can be classified as TK⁺ or as TK⁻. Secondly, their chromosomes can be examined cytologically and a list made of which human chromosomes are present in which progeny clones. Comparison of the two kinds of information enables the presence of the human TK^+ gene to be correlated with the presence of an individual human chromosome. In this case, correlation was with chromosome 17, showing that the gene for thymidine kinase is located on chromosome 17 in humans.

Many other genes have subsequently been located on chromosomes by similar techniques, including several genes of medical interest. For example the genes for the α and β chains of human haemoglobin (which may be defective in certain forms of anaemia) have been mapped to chromsomes 16 and 11 respectively. (In the case of the globin genes the technique has to be modified slightly since the globin genes are *not* expressed in these cultures. The progeny clones have, therefore, to be examined for the presence of the human globin gene DNA sequence by the techniques of nucleic acid biochemistry rather than for the presence of the human globin polypeptides.) A gene which confers unusual sensitivity to a virus which causes respiratory defects (coronavirus) has been mapped to human chromosome 15 by these techniques.

A second kind of information which can be derived from the analysis of parasexual cycle progeny clones is whether two genes are on the same chromosome as each other or on two different chromosomes. Consider a hybrid between a mouse cell line which is mutant for two genes ($A^- B^-$) fused to a human cell line which has both genes in the wild type ($A^+ B^+$) form (Fig. 2–6). If A and B happen to be on the same human chromosome, the progeny clones will either have that chromosome and hence be $A^+ B^+$, or not have that chromosome and hence be $A^- B^-$. (Remember that meiosis is not involved, so there is no chiasma formation leading to $A^+ B^-$ or $A^- B^+$ recombinant chromosomes.) In contrast, if A and B are on different human chromosomes, the progeny clones may only have one out of the two human chromosomes involved, and so also may be $A^+ B^-$ or $A^- B^+$. In summary, the resulting progeny are only $A^+ B^+$ and $A^- B^-$ when the two genes are on the same human chromosome, but they are $A^+ B^+$, $A^+ B^-$, $A^- B^+$ and $A^- B^-$ when the two genes are on different human chromosomes. Therefore it is possible to decide if the two genes are on the same chromosome i.e. whether or not they are 'syntenic'.

Deciding that two genes *are* on the same chromosome is not the same thing as deciding how *far apart* these two genes are on the same chromosome. Techniques for doing this via the parasexual cycle have been developed. The best method involves using human cells which have a variety of chromosome breaks, following heavy irradiation, as a parent for the parasexual 'cross'. The further apart two genes are on the same chromosome, the more likely it is that there will be a radiation induced chromosome break between them and, therefore, they will be separated more frequently during the parasexual cycle. The mathematics for estimating their distance apart from the frequency of progeny clones is complex and will not be discussed here.

2.4   Applications of the human gene map

The use of these techniques in conjunction with conventional genetic

(a) Two genes on the same human chromosome

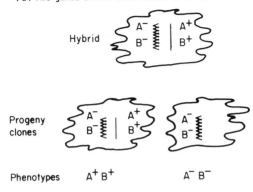

(b) Two genes on different human chromosomes

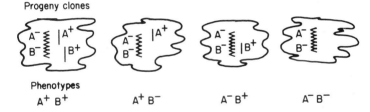

Fig. 2-6 Progeny clones from hybrids containing two human genes $A^+ B^+$ arising from a mouse/human hybrid when the two genes are (a) on the same human chromosome or (b) on different human chromosomes. Human chromsomes are drawn as straight lines, mouse chromsomes as wavy lines. (For simplicity, all chromosomes not carrying genes A or B have been omitted.)

methods has led to a rapid increase in knowledge of the human genetic map. Several hundred human genes have been mapped to chromosomes, including more than 20 to chromosome 1 and more than 100 to the X chromosome.

Information on the human genetic map is of value in the contexts of pure research and medicine. In the former case, for example, information as to whether functionally related genes are clustered gives clues to the way in which activity of the genes may be regulated. In medicine a detailed genetic map is of great value in genetic counselling. Genetic defects may not be recognizable in the amniotic fluid cells obtained by amniocentesis. However, if the gene for a defect is very closely linked to a mutation which *can* be recognized in amniotic fluid cells, the presence of this mutation can be screened for and the probability that the genetic defect is also present can be estimated. Advice can then be given to the mother who is considering abortion.

3 The use of Cell Fusion to produce Pure Antibodies

3.1 Introduction

The use of cell fusion technology took an important step into new areas in 1975 when G. Kohler and C. Milstein announced the fusion of mouse spleen cells with an established tumour cell line. The hybrids which were produced ('hybridomas') secreted a specific type of antibody (a 'monoclonal antibody'). This discovery has had a revolutionary impact in immunology.

Antibodies have, of course, been studied for many years. It has long been known that when foreign substances are present in the blood, the body responds by producing a special group of immunoglobulin proteins known as antibodies. The foreign substances are called antigens. A pure foreign substance in the blood would be a single antigen, whereas a bacterium or a virus would have several antigens on its surface and would therefore elicit several antibodies in response.

In addition to their natural role in the human body, antibodies can be obtained for medical use or for use in biological research by repeated injection of an antigen into an animal such as a rabbit. Such antibodies have two serious disadvantages. Firstly, they are impure because, although the rabbit responds to the injected antigen, it is inevitably producing many other antibodies against other antigens at the same time. The result is a mixture of antibodies which is very difficult to purify. The second disadvantage is that the antibodies are produced in limited quantities. It has long been the dream of immunologists to have ways of producing pure antibodies in large quantities. Kohler and Milstein opened the way to this goal with the production of hybridomas.

3.2 The production of hybridomas

Milstein had been culturing cells of a myeloma, which is a particular kind of cancer. Myeloma cells can be grown as established cell cultures and produce immunoglobulins which are very similar to, and probably identical with, the immunoglobulins which we know as antibodies. It is of particular interest that the myelomas secrete these proteins at a very high rate and they had been studied biochemically for some years as a model system of antibody production. However, the nature of the immunoglobulins secreted by the myeloma cannot be specified.

Milstein and his colleagues conceived the idea of fusing mouse myeloma cells with cells derived from spleens of mice which had been injected with a known antigen (Fig. 3–1). The resulting hybrids are called hybridomas. Individual spleen cells are committed to the synthesis of individual antibodies and therefore contribute information for the synthesis of a single antibody to each hybridoma. The myeloma cells contribute information for synthesis of non-specific immunoglobin and therefore hybridomas originally produced a mixture of the specific antibody and the non-specific immunoglobin. More recently, mutant myeloma cell lines have been isolated which have lost their ability to produce their own immunoglobins but which can still fuse with spleen cells to produce hybridomas which, in this case, only secrete the single antibody specified by the spleen cell. Such single antibodies are called monoclonal antibodies (or McAbs).

The frequency of spleen cell and myeloma fusion is quite low. At best only about 1 in 1000 myeloma cells will give a viable hybrid. Polyethylene glycol (cf. p. 12) is used to increase the frequency and much effort has gone into optimizing the details of fusion conditions by adjusting the ratio of spleen/myeloma cells, the pH, the serum present in the medium etc. Even so, positive selection of the hybridoma from the great excess of unfused spleen and myeloma cells is necessary. This can be done by using mutant myeloma cells which are HGPRT⁻ and the growing in HAT medium (as described previously, p. 12). There is no need to select actively against spleen cells, because they are primary cultures (p. 4) which die off naturally after a while.

Individual spleen cells derived from the mouse are committed to secrete different antibodies in response to different antigens. The mouse from which the spleen cells are derived will inevitably have been exposed to many environmental antigens, and its cells will be committed to produce a range of

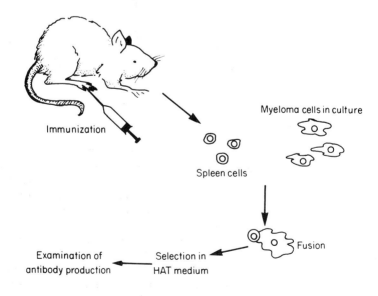

Fig. 3–1   The technique of hybridoma production. See text for details.

antibodies at the start of the experiment. Although a chosen antigen is injected into the animal before the experiment to enrich the proportion of spleen cells committed to produce the corresponding antigen, many other cells will also be present in the spleen cell preparation. Therefore, when the spleen cells are fused with myeloma cells, the resulting hybridomas are a mixture of types, depending on the spleen cell parent of the fusion. Therefore, the hybridomas must be individually checked to pick the one which is producing the required antibody. When the required hybridoma has been identified, it can be cultured indefinitely as an established cell culture.

Unfortunately, antibody production is not always a stable property of hybridomas; some hybridomas can lose the property spontaneously during routine subculture. In some cases this is due to the loss of those chromosomes carrying the genes for antibody production during subculture in a similar way to chromosome loss from other cell hybrids described in Chapter 2. It is possible to overcome this problem by inducing mutations in *other* genes which are in the same chromosome as the genes for immunoglobulin synthesis. Then the hybridoma can be grown in media which favours cells carrying these mutants. This gives positive selection for the drug resistance gene and hence for that chromosome which carries it *and* the gene for antibody production.

When the appropriate hybridoma has been isolated it is then grown either in culture vessels or it can be injected into the body fluids of a mouse (or into a rat, if the myeloma was originally derived from that species). Inside the animal the hybridoma grows as an ascites tumour, (that is as a free living tumour in the body fluids, rather like a suspension culture than as a solid tumour). The advantage of growing the hybridoma within the body of the animal is that the antibody yield is higher; in culture vessels the yield is around 0.01 mg ml^{-1}, whereas the blood of an animal carrying a large number of ascites cells is around $5-20$ mg ml^{-1}. The disadvantage is that the antibody produced in this way is contaminated with other proteins which are naturally present in the blood and it is difficult to purify the specific antibody to a purity of greater than 95%.

3.3 Some uses of monoclonal antibodies

3.3.1 *Monoclonal antibodies against tumour associated antigens*

Certain tumour cells express surface antigens that are not present on the normal cells in the tissue from which they were derived. If monoclonal antibodies are produced which react specifically with these antigens, they will bind specifically to the tumour cell surface. This can be put to several uses.

The tumour specific antibodies can be used as tools to study the spread of cancer cells around the body, either in experimental studies in animals or during clinical treatment of human cancers. For example, the technique has been used to detect the spread of human brain tumour cells to secondary sites elsewhere in the body. The monoclonal antibody which binds to the antigen on the brain tumour cells is visualized by the addition of a dye which binds to antibodies and can be seen as a dark layer around tumour cells.

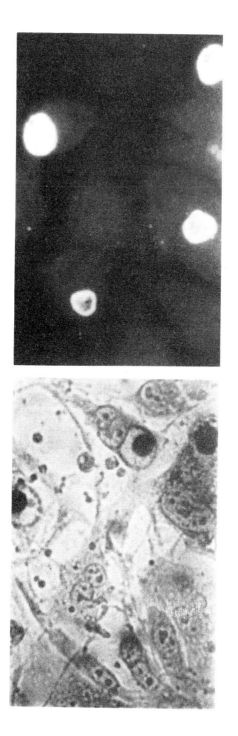

Fig. 3–2 Detection of *Chlamydia* using monoclonal antibodies. *Chlamydia* inclusions have been detected in the cells on the left using iodine stain (dark regions) and on the right using monoclonal antibodies linked to a fluorescent dye (light regions). (Courtesy of Dr N. C. Nowinski, Genetic Systems Corporation, Seattle. Taken from *Science* (1983) volume 219, pages 637–44. Copyright 1983 by the American Association for the Advancement of Science.)

Treatment of cancers using monoclonal antibodies may be attempted in the near future. This could be done by chemically joining a drug to the monoclonal antibody using the specificity of the monoclonal antibody interaction with the tumour cell surface as an effective way of delivering the drug to its target. Alternatively, the monoclonal antibody may have direct effects on the cell surface of the cancer cell, leading to its destruction. Work on animal tumours (for example, mouse leukaemia) has given promising results using this approach, but clear evidence of clinical success in treating human tumours is not yet available. Clearly much interest will be paid to future development in this field of research.

3.3.2 Monoclonal antibodies against rabies virus

Hybridomas have been isolated which produce monoclonal antibodies against a variety of pathogenic viruses and bacteria. One particularly important example is the development of monoclonal antibodies against rabies virus. These have been shown to protect mice against viral infection and have considerable potential for use in humans. The situation is slightly more complicated than may appear at first sight however, because different strains of rabies virus have different antigens and therefore are recognized by different antibodies. The monoclonal antibody must therefore be specific for the strain of rabies virus.

3.3.3 Monoclonal antibodies for diagnosis of infectious diseases in humans

Rapid diagnosis of the nature of infection in human disease is often of critical importance in devising the optimal strategy for treatment. In some diseases the use of monoclonal antibodies is emerging as an effective alternative to conventional microscopic or microbiological identification of microorganisms.

For example, monoclonal antibodies can be used for rapid diagnosis of the causative agent for human sexually transmitted diseases, which can include the bacterium *Neisseria gonorrhoeae*, the intracellular pathogen *Chlamydia trachomatis* and certain herpes simplex viruses. *Neisseria* and *Chlamydia* infections often present very similar clinical symptoms and are particularly difficult to distinguish. However, it has been found that the use of monoclonal antibodies provides a rapid and clear test for these organisms (Fig. 3–2).

4 Uptake of Foreign DNA into Mammalian Cells

4.1 Introduction

Geneticists have known for many years that DNA derived from one strain of bacteria may be taken up by another strain and stably incorporated into its genetic material. This phenomenon is called transformation (see section 1.2). For example, if DNA is extracted from a bacterium which carries a gene for resistance to a particular antibiotic, and is added to a culture of cells which do not carry the gene for resistance, some recipient cells become resistant due to the incorporation of a fragment of the donor's DNA into the recipient's chromosome.

This technique has been widely used to study the arrangement of genes on the bacterial 'chromosomes'. Its main use is to demonstrate whether or not genes are very close to each other on the donor chromosome. The underlying rationale is as follows. The DNA used for transformation is in small pieces. Usually only one fragment of DNA goes into an individual recipient cell and so that recipient cell only receives the genes on a single small piece of DNA. If two genes are located *far apart* on the donor chromosome, they will be on different fragments of DNA and it is therefore very unlikely that an individual recipient will receive both genes. Individual recipient cells (and their progeny) will be transformed for one gene *or* the other. In contrast, if two genes are very close together on the donor chromosome, they will often be on the same small fragment of transforming DNA and will often *both* go into the same recipient cell. This is termed 'co-transformation'. The closer the two genes are on the donor chromosome, the more often they will be on the same DNA fragment and hence the higher the frequency of co-transformation. Co-transformation is therefore used to establish the distance between closely linked genes in bacteria.

Can similar experimental techniques be developed for mammalian cell cultures? If so, it would enable genetic analysis of small regions of human (or other mammals') chromosomes by measurement of co-transformation frequencies. There are also further implications in the long term. If transformation could be carried out on embryos or gametes, it might allow the introduction of new DNA into the genetic material of domesticated animals, modifying their characteristics in ways that conventional breeding and selection methods are unable to do. A more controversial possibility is that a cure for human genetic diseases might be possible if transformation was carried out on cells of individuals carrying genetic defects. In this case, treatment of genetic disease would be attempted by 'gene therapy' rather than by trying to rectify the patients symptoms, which are caused by the lack of a correct gene copy.

Possible ways of introducing DNA into mammalian cells in culture will be discussed in this chapter, and the introduction of DNA into mammalian embryos in Chapter 6.

4.2   Uptake of chromosomes into mammalian cells

DNA in the bacterial 'chromosome' is in the form of a circular DNA molecule. In contrast, the DNA of eukaryotic chromosomes is complexed with histones and other proteins, and at mitosis it becomes packed into well structured, rod-shaped chromosomes. Attempts at transformation in mammalian cells could therefore involve the transfer of chromosomal material from donor to recipient, or they could involve transfer of purified DNA from donor to recipient. Both of these methods have been successfully carried out.

A typical experiment for the transfer of genetic information via chromosomal material is shown in Fig. 4–1. Donor cells are prepared for the experiment by culturing in medium containing colchicine, a drug which blocks their progress around the cell cycle in mitosis. Their DNA and protein are therefore condensed in the typical form of mitotic chromosomes, rather than being in the more diffuse state which is found in interphase nuclei. The cells are disrupted and the chromosomes are isolated by means of differential centrifugation. The chromosomes are added to a recipient culture, and some of them (around 2 per cell) are engulfed and taken into the cytoplasm by phagocytosis (Fig. 4–2). These engulfed chromsomes are broken into smaller fragments in the cytoplasm and these fragments are mostly completely degraded. However, electron microscopy suggests that occasional small chromosome fragments do enter the nucleus. This may result in transformation of the recipient cell.

A selective technique is needed to pick out the few transformed cells from the great majority of recipients which do not become transformed. The $HGPRT^+$ gene is an example of a suitable selectable marker (see p. 12). It will be recalled that $HGPRT^+$ cells can grow in HAT medium but $HGPRT^-$ cells cannot. After chromosomes from an $HGPRT^+$ donor are added to an $HGPRT^-$ recipient, the recipient cells may be plated in HAT medium (Fig. 4–1). Recipient cells which have been transformed from $HGPRT^-$ to $HGPRT^+$ will grow into colonies and can be subcultured for further study.

The incorporation of foreign chromosomal fragments into the nucleus appears to be in two stages. In the first stage, a fragment of the foreign chromosome is replicated in the nucleus, but is not incorporated into the chromosome. These fragments do not have a centromere and therefore cannot attach to the mitotic spindle. Consequently, they do not move to the poles during mitosis and many of the fragments get lost during the mitotic divisions which follow chromosome uptake. This correlates with the observation that the $HGPRT^+$ phenotype is unstable during this phase and may revert to $HGPRT^-$.

In the second stage, the $HGPRT^+$ phenotype becomes stable, following the integration of the foreign chromosomal fragment into one of the chromosomes of the recipient cell. The incoming gene is now distributed at mitosis in a

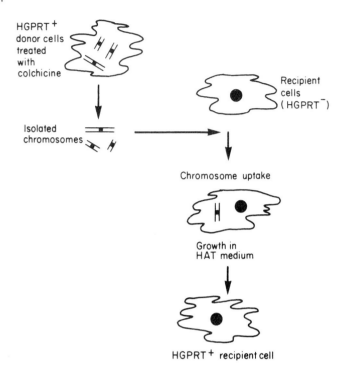

Fig. 4–1 A diagrammatic representation of transfer of genetic information from one cell to another using isolated chromosomes. In this experiment, selection is being made for the HGPRT⁺ gene. For fuller explanation, see text.

regular fashion, like the other genes in the cell. Integration occurs at random sites on the chromosomes of the recipient cell, rather than just at the original HGPRT gene locus.

Optimal conditions for transformation include treatment of the donor chromosomes with calcium phosphate and the recipient cells with dimethylsulphoxide. The efficiency of transformation varies considerably between recipient cell lines, but is usually quite low (around 2 in 100 000 recipient cells incorporate a specific gene).

The size of the chromosomal fragment which becomes integrated is small (only a few per cent of a chromosome). This means that this type of chromosome-mediated transformation can be used to investigate whether or not two genes are close together on the same chromosome. The argument is the same as has been outlined previously for bacterial transformation in section 4.1. If two genes are close together they will tend to be on the same small transforming fragment of chromosome and so the two genes will be transformed. The frequency of co-transformation can therefore be used to

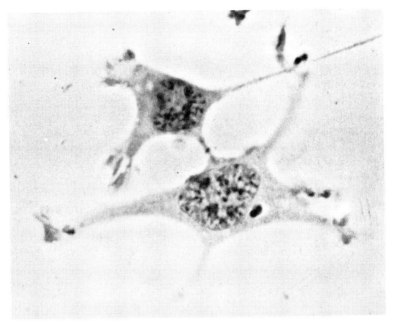

Fig. 4–2 Mouse cells growing in culture after addition of isolated chromosomes to the growth medium. One cell has incorporated a chromosome into its cytoplasm, this appears as a dark region. The cells have been stained with Feulgen stain. (Courtesy of Dr O. Wesley McBride, National Cancer Institute, Bethesda, U.S.A. and the Editor of Stadler Genetics Symposium.)

assess the proximity of two genes on a human chromosome. This technique may provide more precise information about the distances between pairs of human genes than can be obtained by the parasexual cycle discussed in Chapter 2.

4.3 Uptake of DNA into mammalian cells

The chromosome-mediated gene transfer technique outlined in the preceding section has a disadvantage. Transformed cells are in a small minority and a selective technique (e.g. the selection of HGPRT$^+$ cells by growth in HAT medium) has to be used to pick out the transformants. However, suitable selective techniques only exist for a few genes; many do *not* confer an obvious selective advantage on cells in culture and therefore are impossible to detect by this method. For example, the gene for β globin (part of the haemoglobin molecule) does *not* confer a selective advantage, but studies of transformation of this gene would be of potential medical and scientific interest. In the former context, it might be possible to cure some human hereditary anaemias which involve defects of this gene by inserting normal gene copies into cells (see also Chapter 6.) In the context of pure science, it would be interesting to molecular biologists to be able to alter DNA bases in and around the globin gene and subsequently to reinsert the modified gene into a chromosome. The effects of altering individual bases on the pattern of gene expression could be studied. This may provide a rather direct approach to the study of molecular basis of control of gene activity in eukaryotes.

Transformation of non-selectable genes can be achieved using purified DNA (rather than whole chromosomes). This technique is rather similar to that used for bacterial transformation. DNA is extracted from donor cells precipitated with calcium salts and added to the recipient cells. Transformation is only observed at a low frequency (around 1 in 100 000 recipient cells). However, the situation is not what it seems at first sight. As in bacteria, only a minority of cells are capable of being transformed; these cells are said to be 'competent' for transformation. Altough the rate of transformation in the population as a whole is low, the rate of transformation in this small sub-population of cells is high, and cells which are transformed by one gene are frequently transformed by a second gene on a *separate* piece of DNA. This phenomenon can be used to pick out cells which are transformed by genes which do not confer selectable properties on recipient cells.

M. Wigler carried out one of the earliest experiments in this field in 1979 (Fig. 4–3), working with the TK$^+$ gene. Cells which are TK$^+$ can be selected from TK$^-$ cells in HAT medium (see p. 12). DNA which carried the TK$^+$ gene was isolated from a virus using molecular biological techniques and when this DNA was added to TK$^-$ cells, TK$^+$ transformants could be detected. Wigler then added a mixture of the TK$^+$ gene DNA and rabbit globin gene DNA to the TK$^-$ cells and carried out selection for TK$^+$ on HAT medium. He found that most of the TK$^+$ colonies which grew had not only incorporated the TK$^+$ gene but had also incorporated the rabbit globin gene into their chromosomes. The presence of the rabbit globin gene DNA in the transformed cells was demonstrated by a molecular biological technique which detects specific DNA base sequences. In outline, the recipient cell DNA is converted from a double helix to a single stranded form and then mixed with a radioactive sample of authentic globin gene DNA, known as a probe. Under certain conditions, single strands of the probe DNA reform double helical molecules with single strands of any cellular DNA which has a complementary base sequence. Thus the hybridization of the radioactive probe to the cellular DNA demonstrates the presence of the β globin gene in the transformants.

The globin genes which have been inserted into the recipient cells are stably inserted during subsequent divisions. There may be one or several copies of the incoming gene in an individual recipient cell and its progeny. Many other kinds of DNA apart from the β globin gene have been inserted into recipient cells and the technique appears to be of general applicability. Recent work involves micro-injection of very small volumes of liquid containing genes into the nuclei of cells grown in culture or growing in embryos to bring about transformation. The possible long-term medical implications of the development of these techniques are considered in Chapter 6.

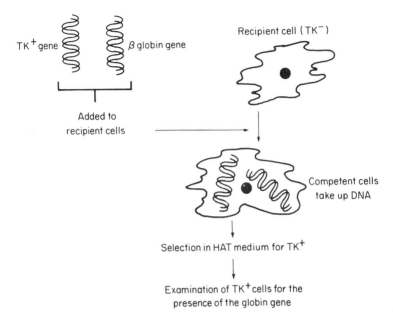

Fig. 4–3 A diagrammatic representation of the simultaneous uptake of DNA carrying a selectable marker (the TK^+ gene) and DNA carrying an unselectable marker (the β globin gene) into competent mammalian cells. See text for further details.

4.4   The use of transformation in cancer research

The ability to introduce DNA into mammalian cells offers a valuable opportunity to examine the consequences of introducing a specific DNA sequence into cells. An interesting and important example of this type of work has recently emerged in the area of cancer research.

A fundamental question in cancer research is 'What is the nature of the genetic changes which occur during cancer development?' Chemicals which are known to induce mutations nearly always cause cancer and this has led to the widely held view that changes in DNA play a major role in carcinogenesis. However, there is dispute over the nature of these changes. One school of thought suggests that they may involve the substitution of one base in a normal gene by a different base, but other changes in the DNA are also possible. There is indirect evidence that suggests that a gene may be moved to a different position on the chromosome (or on another chromosome) during cancer induction. It is known that base sequences flanking all genes can affect their level of expression (i.e. the rate at which they are transcribed into messenger RNA and hence code for protein). If a gene is moved to a different position (and such movements are, surprisingly, now known to occur in cells) it will have different flanking sequences and its level of expression might well change. If the gene specified a regulator of cell division, altered levels of expression could alter the control of cell division.

A direct approach to this question is currently being made by studying the consequences of the uptake of DNA carrying cancer genes or closely related base sequences into cells grown in culture. The basic strategy is to see which types of DNA convert the cells to cancerous growth. Fortunately, this can be studied in culture because cancerous cells grow as a piled up heap in contrast to the normal single layer (Fig. 4–4). If DNA uptake leads to cells becoming cancerous, this can therefore be seen by direct observation of the cultures. Using this as an assay, it is possible to identify the DNA in tumours which is responsible for cancerous growth.

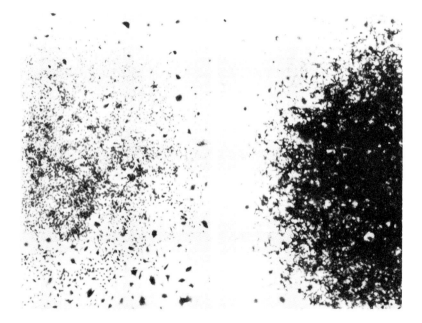

Fig. 4–4   Part of a normal colony of hamster cells (*left*) and part of a colony of cancerous hamster cells (*right*). The cancerous cells grow in a more 'piled up' arrangement and appear darker when viewed from above. (Courtesy of Dr R. F. Newbold, Pollards Wood Research Station, Royal Cancer Hospital and the Editors of *Nature*.)

C.J. Tabin has isolated a gene from human bladder cancer which causes cancerous growth when taken up by recipient cells grown in culture. A very similar DNA sequence is present in normal bladder cells but this does *not* alter the growth form when taken up by the recipient cells. What is the precise difference between the mutant cancer gene and its normal equivalent? Tabin and his colleagues at the Massachusetts Institute of Technology have shown that hundreds of bases are identical in the DNA sequence of the cancer gene and its normal equivalent but a single base has changed. One guanine is replaced by a thymine which changes a coding triplet of bases from GGC to GTC. The consequence of this is that a single amino acid in the protein which is coded by the gene is changed from a glycine in normal cells to a valine in cancer cells. This amino acid substitution is likely to alter the three dimensional folding of the protein, but it is not clear at the present time why it leads to cancerous growth.

This experiment indicates that, in this case, the simplest kind of mutation (similar to that known to occur in some hereditary diseases, such as sickle cell anaemia) has led to cancerous growth. However, there are several reservations that should be borne in mind when interpreting these results. The workers certainly do not suggest that *all* cancers involve the same base change. Secondly, the *in vitro* study of cancerous growth is probably only equivalent to one of several steps in the development of a malignant tumour. (For example, acquisition of the ability of cancer cells to migrate to new sites within the body to form secondary growths probably represents a different step.) Bearing these reservations in mind, however, the development of the type of experiment outlined here does represent a very significant step forward in cancer research.

5   Gene Cloning in Mammalian Cells

5.1   Gene cloning in bacteria

One of the most dramatic advances in biology in recent years has been the development of techniques to introduce new genetic material into bacterial cells by gene cloning. In gene cloning experiments genes are carried on vectors which may be autonomously replicating small circles of bacterial DNA called plasmids, or they may be the DNA of the 'chromosome' of a phage (i.e. bacterial virus). The genes may be derived from other bacterial cells, from eukaryotic cells, or they may be chemically synthesized. When a piece of foreign DNA has been inserted into a vector, the new combination of DNA is called a recombinant DNA molecule.

In some cases, when the foreign genes have been introduced in a vector into bacterial cells, they are transcribed into mRNA and the proteins which are coded for by the foreign genes are produced. If such expression of inserted genes does occur, the gene products can then be obtained from the bacterial cells. For example, the human insulin gene has been cloned in bacteria and the insulin which is produced in large quantities can be used in the treatment of human diabetes, in place of the pancreatic insulin conventionally derived from cattle and pigs.

It is not within the scope of this book to provide a general account of gene cloning in bacteria, and the reader is referred to other texts for a fuller discussion (see, for example Old and Primrose, 1981). However, a brief outline is presented here as background information for the discussion of gene cloning in mammalian cells which follows, and for the discussion of genetic engineering in plants (Chapter 7).

In vitro insertion of pieces of foreign DNA into plasmids or viral DNA is largely dependent on a group of bacterial enzymes called restriction endonucleases. The biological role of these restriction enzymes is to cut DNA at specific base sequences (which vary with different enzymes) and so degrade incoming foreign (viral) DNA. Bacteria protect their own DNA by adding methyl groups to bases at these specific base sequences using a second group of enzymes. The methylated DNA is not degraded by restriction endonucleases.

Restriction endonucleases may be used in the laboratory in several different ways to construct recombinant DNA molecules. Some restriction endonucleases make staggered cuts at particular sequences of DNA (regardless of the origin of that DNA) and these staggered cuts are useful in rejoining different DNA fragments. The staggered cut produced by one such enzyme Eco R1 is shown in Fig. 5–1. (Restriction endonucleases are named after their bacterial source; Eco R1 is derived from *Escherichia coli*.) If the same enzyme is used on DNA from two different sources, it will cut them both at the same sequence, producing the same single strand tails on both molecules.

Under suitable conditions, base pairing (i.e. A with T, G with C) will occur between the free single stranded regions because they have complementary base sequences (Fig. 5–1d). The gaps in the sugar phosphate backbone of the DNA molecule between the two pieces of DNA can now be covalently joined using another enzyme (a DNA ligase) to produce a recombinant DNA molecule which contains DNA from two sources.

In the process of gene cloning, one molecule (represented as a circle in Fig. 5–2) would be either plasmid DNA or the DNA of the viral 'chromosome' and the other molecule (represented as a straight line) would be a fragment of foreign DNA which is being inserted. The restriction enzyme cuts open the circle as described above (Fig. 5–1) leaving two single stranded tails. These base pair with the single stranded tails of the foreign DNA which is being inserted. If a plasmid is used, some of the plasmid DNA molecules are taken up

(a) Target sequence

```
        —GAATTC—              wwww GAATTC wwww
        —CTTAAG—              wwww CTTAAG wwww
```

(b) Sites of cuts

```
              ↓                          ↓
        —GAATTC              wwww GAATTC wwww
        —CTTAAG              wwwwCTTAAG wwww
                  ↑                      ↑
(c) Resulting breaks
```

```
     —G        AATTC—        wwww C        AATTC wwww
     —CTTAA       G—         wwwwCTTAA        G wwww
```

(d) Base pairing between DNA from different sources

```
              — GAATTC wwww
              — CTTAAG wwww
```

Fig. 5–1   The use of Eco R1 to produce staggered cuts in DNA. (a) The target sequence for the enzyme is a specific sequence of six bases. Bases which are outside the target sequence are represented as lines in this figure. In order to distinguish between DNA from two sources, DNA on the left of the figure is represented by straight lines and DNA on the right is represented by wavy lines. (b) The sites of the cuts by the enzyme are represented by vertical arrows. (c) The resulting break produces single stranded tails. (d) Base pairing can occur between DNA from different sources, because they have single strand tails with complementary base sequences.

into bacterial cells following calcium chloride treatment and heat shock at 42°C and then replicate independently of the chromosome. Plasmids used for gene cloning often carry genes for drug resistance and it is therefore possible to pick out those bacteria which have received the plasmids by plating the bacterial cells on medium containing the drug. In the case of the viral 'chromosome', the DNA can be packaged into viral protein coats and injected into bacterial cells

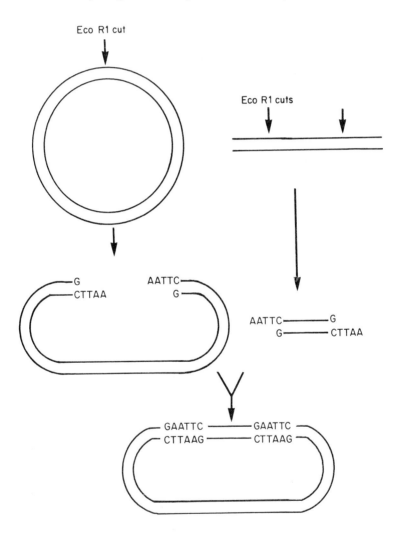

Fig. 5–2   Insertion of a fragment of foreign DNA into a plasmid vector. Both DNA molecules have been cut by the same restriction enzyme and so have complementary single strand tails. Bases are only shown at the restriction site, other regions of the DNA molecule are represented simply as lines.

provided that the recombinant DNA molecule is not too large to be packaged in the protein coat. Some non-essential regions of the phage DNA may be removed to ensure that the recombinant molecule, containing 'extra' DNA, is a suitable size.

Techniques of this kind have revolutionized bacterial genetics. Is it possible to use analogous techniques in eukaryotic cells? There are several good reasons why one may wish to achieve this. Firstly, at the molecular level, gene expression in eukaryotic cells has fundamental differences from that of prokaryotic cells. Therefore, studies on the control of gene expression of eukaryotic genes are more meaningful if the genes are cloned in eukaryotic rather than prokaryotic cells.

A second reason for wishing to clone genes in eukaryotic cells is that these cells modify some proteins *after* synthesis in ways which are not carried out in bacteria. In particular, eukaryotic proteins may be glycosylated, that is they may have sugar residues linked to some of their amino acids. For example, the gene for β interferon (a promising antiviral agent) has been cloned in bacteria but the interferon produced in these cells has not got the normal pattern of glycosylation and this alters it properties. Cloning the β interferon gene in mammalian cells however gives a fully glycosylated protein molecule. Such considerations may be very important in the commercial production of certain proteins.

A third reason for interest in gene cloning in eukaryotic cells is to develop techniques for the introduction of new kinds of genetic material into whole organisms. This is discussed in respect of animals in Chapter 6 and in respect of plants in Chapter 7.

5.2   Uptake of bacterial plasmids into mammalian cells

Bacterial cells occasionally fuse with mammalian cells when they are mixed together. The spontaneous frequency is very low but can be increased by removing the bacterial cell wall (i.e. by making bacterial protoplasts) and by adding polyethylene glycol. It has recently been shown that all mammalian cells present are involved in fusions if bacterial protoplasts are in great excess.

This provides a direct way to introduce cloned genes into mammalian cells. Fusion of bacterial protoplasts, made from a bacterial strain which carries a plasmid which has an inserted gene, with mammalian cells results in the presence of the plasmid in the mammalian cell. This allows the study of the expression of cloned genes in mammalian cells. However, the bacterial plasmid is only replicated in appropriate bacteria and is not replicated in mammalian cells and so, as the mammalian cells divide, the plasmid is rapidly lost from the population. This technique is obviously not suitable for long-term experiments.

5.3   Viral vectors for gene cloning in mammalian cells

Plasmids and viral DNA are used as vectors for gene cloning in bacteria and so molecular biologists have naturally considered whether any endogenous plasmids or viruses exist which may serve as vectors in mammalian cells. Unfortunately, endogenous plasmids have not been recognized in higher eukaryotic cells, but several viruses have been developed as cloning vectors. One of the most widely studied of these is SV40 (Simian Virus 40).

When SV40 infects mammalian cells growing in culture it undergoes two distinctly different series of events depending on the nature of the cells. When added to certain monkey cell lines ('permissive hosts') it undergoes a conventional viral life cycle leading to cell lysis and death. At the start of the infection the viral genetic material enters the nucleus and some of its genes (the 'early genes') are expressed. This leads to the commencement of viral DNA replication after about eight hours. Subsequently other viral genes ('late genes') are expressed which specify viral structural proteins. After about 36 hours progeny virus particles are liberated and the cell is destroyed. This sequence of events is called the lytic cycle.

However, if SV40 infects mouse or rat cells in culture ('non-permissive hosts') the lytic cycle does not occur, because the cells are not able to support viral DNA replication. The great majority of the mouse or rat cells are therefore not affected by the viral infection. Occasional cells, however, incorporate the SV40 genetic material into their chromosomes leading to the cancerous state (see p. 28 and Fig. 4–4).

The molecular biology of SV40 has been studied intensively for a number of years and this makes it an obvious choice for development as a cloning vector for mammalian cells. The viral genome consists of a single circular double stranded DNA molecule, 5243 base pairs in length. The entire base sequence is known and there are several useful sites at which restriction endonucleases can cut open the circle prior to the insertion of a fragment of foreign DNA.

SV40 can be used as a cloning vector in non-permissive host cells or in permissive host cells. In the former case, foreign genes are inserted into the viral chromosome *in vitro* and are then incorporated into a mammalian cell chromosome along with SV40 DNA during infection. Initially this approach seemed promising, because the foreign DNA replicates with the chromosome, but now there are reservations. One of the problems is that the integration of SV40 DNA is accompanied by a complex series of chromosome rearrrangements. This makes it impossible to know how the cloned genes are arranged, and makes it difficult to interpret studies on their gene expression. Work has therefore been concentrated on SV40 cloning in permissive hosts.

If foreign genes are inserted into SV40 *in vitro* and a permissive host is infected, the inserted gene replicates along with the rest of the viral genome to produce around 10 000 or 100 000 copies. Since the hybrid DNA molecule (SV40 plus foreign DNA) has to be packaged into a protein coat for infection, there is a limit to the total size of the hybrid DNA molecule which can be used (cf. page 33). It is therefore necessary to remove some genes from the SV40 chromosome to make room for the foreign DNA. Since the SV40 is consequently lacking in some genes, a simultaneous infection has to be made with a 'helper' virus whose function is to compensate for the deficiencies in the principal virus.

When the foreign gene is inserted at a suitable position on the SV40 genome, it is expressed along with the SV40 genes during infection of permissive host cells. This system has been used to study the expression of many genes in mammalian cells, including the rabbit globin and the rat proinsulin genes. It does, of course, have one serious disadvantage: the host cells are killed during the course of infection. This has led several workers to attempt to produce a 'synthetic' vector which will replicate in mammalian cells without killing them.

5.4   Synthetic vectors for gene cloning in mammalian cells

R.C. Mulligan and P. Berg of Stanford University, California, have constructed a series of vectors, each of which contains fragments of DNA from the SV40 genome, from a bacterial chromosome and also from a bacterial plasmid. The net result is vectors which will replicate either in mammalian cells or bacteria without killing either, and which carry genes that enable them to be selected for in either mammalian cells or bacteria. The component parts of one such vector are considered in Fig. 5–3.

Replication of DNA molecules within cells can only start from specific base sequences called origins of DNA replication. The SV40 origin is therefore present in the vector to allow it to replicate in a permissive host. Certain genes which are also necessary for SV40 DNA replication are present, but genes required for the rest of the SV40 lytic cycle are absent and hence the host cells are not destroyed.

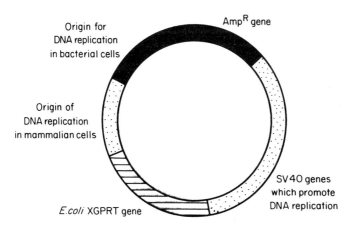

Fig. 5–3   A synthetic cloning vector for mammalian cells constructed by Mulligan and Berg. Stippled regions are derived from SV40. Dark regions are derived from the bacterial plasmid pBR322. Hatched regions are derived from the *Escherichia coli* chromosome.

A gene from the bacterium *Escherichia coli* coding for the enzyme xanthine guanine phosphoribosyl transferase (XGPRT) is present in the vector. This is to provide a way of selecting for the uptake of the vector into mammalian cells. Although this enzyme is slightly different from the mammalian enzyme HGPRT discussed previously (p. 9), it carries out a similar function and if the host cell culture is HGPRT$^-$, those cells which have taken up the vector (including the XGPRT$^+$ gene) are converted from HGPRT$^-$ to HGPRT$^+$. This provides a means of positive selection for cells carrying the vector, simply by plating in HAT medium (p. 12).

The final segment of the vector is derived from a bacterial plasmid pBR322. This has two features. There is an origin of DNA replication which will function in bacteria, enabling the vector to be maintained in bacteria like an ordinary bacterial plasmid. There is also a gene for ampicillin resistance which enables the presence of the vector in bacterial cells to be easily selected for by plating on medium containing this drug.

Overall this provides a vector with considerable technical advantages over the basic SV40 chromosome. It can be maintained in bacteria between times when it is being manipulated (e.g. having DNA sequences inserted or altered). When it is required to study expression of the inserted genes or to obtain the protein coded for by them, the vector can be readily transferred to mammalian cells and maintained there without destroying them. The development of vectors of this kind means that gene cloning in mammalian cells is entering a very fruitful phase, equivalent to that which gene cloning in bacterial cells entered a few years ago.

6   Incorporation of Genes into Whole Animals

6.1   Introduction

Techniques for introducing genetic material into mammalian cells growing in culture have been described in the two preceding chapters. Is it possible to adapt these techniques to introduce new genetic material into whole organisms? There are several general reasons why this would be of interest.

The mechanisms which underlie the differentiation of cells into tissues and organs represent some of the most intriguing and poorly understood problems in biology today. Many genes are only expressed at particular stages in development and in certain tissues. It is obviously not appropriate to analyse these problems in cell cultures. However, if genetic material could be introduced into whole organisms it would be possible to introduce mutant genetic material or chemically modified genetic material into an organism and then to observe the consequences for gene expression during development. Particular base sequences could then be correlated with patterns of gene activity in the whole organism.

Geneticists are also interested in introducing genes into whole animals for quite different reasons. It may provide a useful technique to improve the genetic constitution of farm animals. For example, if it was possible to introduce a gene which led to increased production of growth hormone, this may lead to faster growing animals and would be welcomed by farmers. There is potential here for a whole new dimension to be added to animal breeding.

If techniques become available to introduce genes into mammals, would it be possible to use similar techniques to introduce genes into humans? There are over 2000 human diseases with a genetic basis and it may be possible to treat some of them by gene therapy rather than treating the symptoms which arise from the defective gene. It is possible to envisage two fundamentally different types of gene therapy. In the first case the aim might be to insert 'correct' genes into somatic tissues of a patient, with no effects on his or her offspring. In the second case, the strategy might be to insert genes into an embryo and then reimplant the foetus in the mother. Doctors are currently developing techniques of *in vitro* fertilization and embryo culture to overcome certain forms of human infertility. The development of these techniques would also have implications if gene therapy were ever contemplated.

Several different approaches to the introduction of genetic material into whole organisms are being studied and some of the most significant will be outlined in this chapter. Research in this area has considerable moral and ethical implications which it will be essential for individuals, communities and governments to discuss in the future. It is not within the scope of this book to discuss these but it is hoped to provide some factual basis for these important discussions.

6.2   Replacement of bone marrow cells in animals

Certain types of cells (e.g. bone marrow cells) can be withdrawn from an animal's body, grown in culture and then replaced in the body later. Is it possible to use techniques similar to those described in Chapter 4 to introduce foreign DNA into these cells while they are out of the body in order subsequently to modify the genes of some somatic cells?

In 1980, M.J. Cline, working at the University of California, reported an experiment of this type (Fig. 6–1). Bone marrow cells were withdrawn from a mouse and exposed to DNA containing a gene specifying resistance to an antitumour drug called methotrexate. Some of the cells took up the DNA and consequently became methotrexate resistant. The cells were returned to the body and the animal was given doses of methotrexate for a while (in order to give a selective advantage to the methotrexate resistant cells in a mixture of resistants and sensitives). It was found that these mice were able to tolerate higher levels of methotrexate than control mice whose bone marrow cells had not received the resistance gene DNA.

This experiment has important medical implications. The treatment of human cancer sufferers with drugs like methotrexate often has to be stopped because the drug is killing too many essential bone marrow cells before it has killed all the cancer cells in the body. Might it be possible to remove marrow cells from the patient *before* treatment with the drug, convert them to drug resistance, and then return them to the body? The patient might then be able to tolerate higher (more effective) concentrations of the antitumour drug with less harmful side effects.

Cline was aware of another potential medical implication of his experiment. Several serious hereditary anaemias involve abnormalities of the globin genes. One such disease is beta-zero thalassaemia. Cline speculated that if it was possible to insert a normal globin gene into the marrow cells of a patient and then replace the cells in the patient's body, there may be medical benefit. The medical ethics panel of his university in California were not convinced that the time was ripe to attempt this treatment and refused to give their approval. There was concern that animal experiments had not yet given sufficient evidence that the globin gene would be expressed at a useful level in the patient and also that Cline proposed to insert the gene in the form of a recombinant DNA molecule. The globin gene was to be linked by recombinant DNA techniques to a gene for viral thymidine kinase because this viral gene is more efficient that its mammalian counterpart and may therefore confer a selective advantage on cells which carry it with the associated globin gene. However, the U.S.A. authorities would not approve the use of *any* recombinant DNA molecules for human treatment. In order to circumvent restrictions in his own country, he attempted the experimentation therapy (without any success) on patients in Israel and in Italy and when the situation became public, controversy flared. The U.S. National Institutes of Health, in an unusual

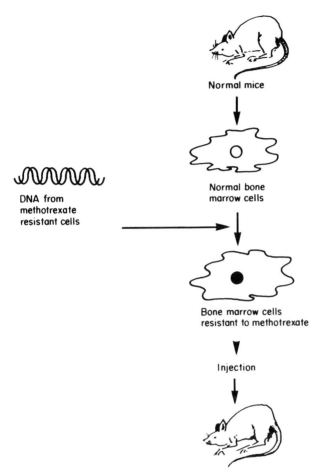

Normal mice

Normal bone
marrow cells

DNA from
methotrexate
resistant cells

Bone marrow cells
resistant to methotrexate

Injection

Fig. 6–1 Diagrammatic representation of an experiment performed by Cline to introduce a gene for methotrexate resistance into bone marrow cells *in vitro* and then to return the cells to the animal.

move, clearly intended to warn other scientists not to ignore restrictions laid on them, withdrew financial support for his work in this area. Currently, other groups are continuing with animal experiments and it seems likely that a fairly similar type of human gene therapy will be attempted somewhere in the world in the next few years.

6.3   Incorporation of cultured cells into embryos

The second approach to introducing foreign genes into whole animals is fundamentally different from the method tried by Cline because it involves working with embryos. Developmental biologists such as B. Mintz have observed that if certain mouse tumour cells called teratocarcinomas are injected into early mouse embyos (at the blastocyst stage) some of the teratocarcinoma cells become incorporated into the embryo as an integral part of it. If the embryo is reimplanted in a foster mother, the resulting 'progeny' mice are mosaics, consisting mostly of tissues derived from the original blastocyst cells, but with some regions produced from the injected teratocarcinoma cells (Fig. 6–2). (The different origins can be recognized because the two cell types carry different genetic markers. The most obvious one is coat colour. The teratocarcinoma cells may be originally derived from an agouti colour mouse and the blastocyst from homozygous black coat colour parents. The resulting mosaic is black with agouti patches.) Although the teratocarcinoma is a cancer cell, interestingly the mosaic progeny are not cancer bearing, implying reversal of the expression of the cancer phenotype.

Occasionally an injected teratocarcinoma cell will be incorporated into the gonads of the developing mosaic and this adds an extra dimension to the situation. When this happens, gametes may be produced which are derived from the teratocarcinoma cell line. If such mosaics are mated with normal mice, their offspring and subsequent generations inherit genes originally derived from the carcinoma in a normal Mendelian fashion.

Developmental geneticists would like to extend this work further. It should be possible to insert genes (which have been previously manipulated in whatever way is required using recombinant DNA technology) into the teratocarcinoma cells, using techniques described in the previous chapter. After the teratocarcinoma cells have been incorporated into mice and their progeny it would be possible to study the effects of any kind of mutation on developmental processes in the whole animal.

The teratocarcinoma system is only applicable to the mouse at present. Conceivably a similar experimental system might be developed in the future for other animals of agricultural interest, providing a way to introduce new genes into livestock. It will be important, and very difficult, to ensure that the new genes which are introduced into animals are expressed correctly. The ways in which the activity of genes is regulated during development is very poorly understood. To obtain correct regulation it is possible that some genes may have to either be inserted at a specific chromosomal location or at least with extensive correct flanking sequences of bases.

6.4   Insertion of DNA into embryos using viruses

Several groups of workers have investigated the possibility that new genetic material may be introduced into embryos using viral vectors. R. Jaenisch and his collaborators at Hamburg University have shown that DNA sequences of the Moloney leukaemia virus are stably incorporated into mouse chromosomes following infection of embryos. Infection is achieved either by co-incubating the embryo with other cells which are producing the virus or by micro-injection of the virus into the blastocyst. When 4 to 16 cell stage mouse embyros are

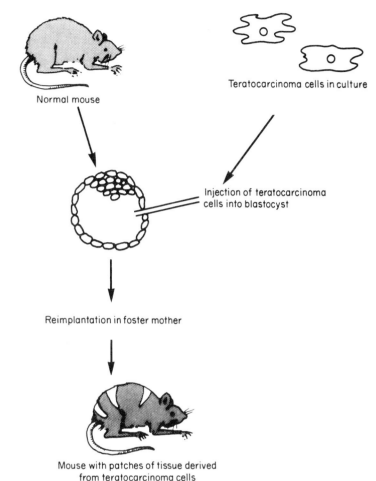

Normal mouse

Teratocarcinoma cells in culture

Injection of teratocarcinoma
cells into blastocyst

Reimplantation in foster mother

Mouse with patches of tissue derived
from teratocarcinoma cells

Fig. 6–2 A diagrammatic representation of an experiment by Mintz involving the injection of teratocarcinoma cells into a mouse embryo. The teratocarcinoma cells and the embryo carry different genes for coat colour. A mosaic mouse results with patches of tissue derived from the teratocarcinoma.

infected and then re-implanted in foster mothers, a high proportion of the resulting mice are mosaics, i.e. animals which are a mixture of cells, some containing a viral DNA sequence integrated into the chromosome and some cells not containing the sequence. This presumably depends on which ancestral cells were infected during the experiment. In some animals, the viral DNA sequence is incorporated into the germ cells and then may be inherited as a normal Mendelian gene in succeeding generations. Thus new genetic material has been introduced into the organism.

It appears possible to extend Jaenisch's work by the use of recombinant DNA techniques to join additional genes to the viral genetic material prior to the infection of embryos. The virus should carry the 'new' genes with it, into the chromosome. The same problem that was mentioned in section 6.3 (p. 40) may arise however, because in Jaenisch's experiment the viral sequences go into different positions on the mouse chromosomes and, in consequence, their expression occurs at different levels in different tissues in various individuals. (The expression of the viral genes in this experiment is estimated by measuring the levels of the RNA sequences which are transcribed from them. This can be done by molecular biological techniques.) It is quite impossible at present to 'direct' the virus to a specific site, which may be necessary to ensure that the genes it carries are expressed at the appropriate level in particular tissues. The present approach can only be to insert genes at random and then see how they are expressed.

6.5 Insertion of DNA into embryos using microinjection

It was noted in section 4.3 that microinjection has made the uptake of DNA into mammalian cells grown in culture much more efficient than simply allowing the cells to take up DNA from the surrounding medium. This direct approach has now been applied to the insertion of genes into fertilized eggs. In 1980, J.W. Gordon and his co-workers at Yale University demonstrated that when DNA sequences were injected into the pronuclei of fertilized mouse oocytes and the oocytes were implanted in foster mothers, two out of seventy eight resulting mice contained the new (injected) DNA sequences, stably incorporated in their chromosomes. As far as they were able to tell, the new DNA sequences were present in all of the cells of these animals, which one might expect as injection is done *before* the egg starts to divide.

In 1981, F. Constantini and E. Lacy at Oxford University demonstrated that not only is injected DNA present in the mice which subsequently develop, but these mice can pass the DNA sequences to their progeny when they are mated with normal mice. They showed this by injecting the rabbit β globin gene (which can be distinguished from its mouse equivalent by molecular biological techniques) into fertilized mouse eggs and studying the resulting mice and then their offspring. E.F. Wagner of the Fox Chase Cancer Centre, Philadelphia and others have extended these observations further by showing that the incorporated rabbit β globin can be used to specify at least some of the rabbit form of the protein within the erythrocytes of the mouse. Together these studies establish the two fundamental points that injected genes can be inherited and can be expressed.

Possible practical benefits from this type of experiment came closer in 1982 when R.D. Palmiter and R.L. Brinster of the University of Washington and Pennsylvania reported a remarkable development in this area. They injected a rat gene coding for growth hormone into fertilized mouse eggs. The experiment differed from earlier ones in that the injected gene was joined by recombinant DNA techniques to another piece of DNA prior to injection. This

DNA fragment was a sequence which originally was adjacent to the mouse metallothionein gene (Fig. 6–3). Metallothionein is a protein which binds to heavy metals and, by doing so, confers resistance to their toxic effects. The interesting point to note about it here is that the level of expression of the gene (i.e. the rate of mRNA production and consequent metallothionein synthesis) can be varied. High levels of zinc in the cell leads to high levels of gene expression. The molecular mechanism of this is not clear, but the important

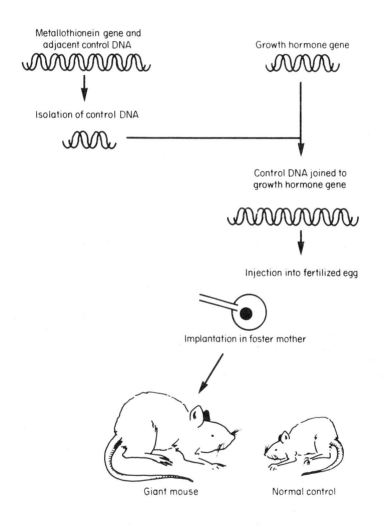

Fig. 6–3   The injection of a growth hormone gene linked to a controlling sequence into mice eggs. See text for further discussion.

point to note is that about 90 bases of DNA adjacent to the gene are essential for the effect of zinc on gene expression.

Palmiter and Brinster joined these 90 bases to the end of the growth hormone gene and then injected the recombinant DNA molecule into fertilized mouse eggs. Some of the resulting mice had incorporated the recombinant DNA molecule into their chromosomes. These mice were fed on a high zinc diet and the zinc apparently interacted with the 90 base pair region and increased the expression of the adjacent gene, with the result that the mice had up to 800 times the normal levels of growth hormone in their tissues. Consequently they grew very rapidly to about twice the normal size (Fig. 6–4).

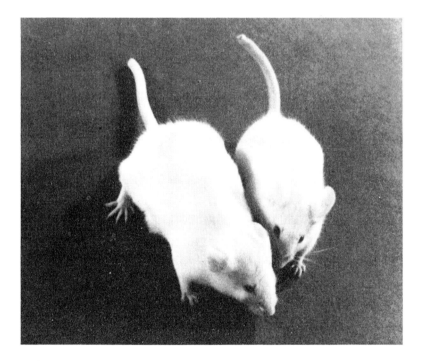

Fig. 6–4 Two mice from a genetic engineering experiment. The mouse on the left contains the rat growth hormone gene fused to mouse metallothionein control DNA. The mouse on the right is a control. (Courtesy of Professor R. L. Brinster, School of Veterinary Medicine, University of Pennsylvania, and the Editor of *Nature*.)

This work has obvious implications as a technology to stimulate rapid growth of commerically valuable animals. Benefit should accrue from shorter production time and possibly from increased efficiency of food utilization. It may be particularly useful to have a gene which can be switched on or off by regulating the zinc levels in the animals, by changes in the zinc content of the diet. High growth hormone levels can also raise milk yield. Apart from its agricultural implications, this technology may also be useful as a model of certain human diseases, such as gigantism, which results from overproduction of growth hormone.

6.6   Future prospects for human gene therapy

Knowledge of insertion of genes into animals is developing very rapidly – Is this leading to the possible use of these techniques for therapy of human genetic diseases in the forseeable future? As mentioned in section 6.1, attempts could be made to insert genes into somatic cells ('patient gene therapy') or to insert genes into fertilized eggs or embryos ('embryo gene therapy').

In the case of patient gene therapy (as attempted by Cline on thalassaemia sufferers, section 6.2), there are fundamental technical problems to overcome. These include the problem of introducing the correct gene into enough cells of the body to improve the patient's condition, the introduction of the gene into a position in the chromosome where its level of expression in regulated normally, and the use of a vector or other method of introducing the gene which does not cause chromosome damage. All these technical problems are serious at the moment but it seems quite likely that they will be solved within about a decade or two. Society will have to judge the ethics of this form of therapy. In the author's view, the ethical problems associated with this form of gene therapy are not fundamentally different from those which arise during the development of a drug therapy. Basically they involve the need to ensure that any experimental therapy does not expose a patient to unnecessary risk compared to likely medical benefit and that the informed consent of the patient should be obtained before any experimental treatment is given.

Embryo gene therapy is much more controversial and has become more widely discussed since gynaecologists have developed *in vitro* fertilization (IVF) techniques to overcome some forms of infertility in women. These techniques involve the fertilization of an egg by sperm in the laboratory and the maintenance of the embryo briefly in culture medium before being implanted in the mother. The development of IVF techniques led to the birth of the first 'test tube baby' in Oldham in the United Kingdom in 1978 and subsequently many other successful pregnancies around the world. It seems likely that these techniques will find widespread application in the future as a treatment for infertility. Is it possible that they may also open the way to the introduction of new genetic material for embryo gene therapy?

Some genetic diseases are caused by a mutation in a *single* Mendelian gene. Examples of this (amongst many others) include sickle cell anaemia, thalassaemias, phenylketonuria, growth hormone deficiency, and Lesch-Nyman syndrome (p. 9). It is estimated that in Britain, serious single gene defects affect just under 1% of births. However many major health hazards, for example, coronary heart disease, hypertension, some psychotic illness, some forms of cancer, have complex causes involving the interaction of *many* different genes and important environmental factors. There is wide agreement that our state of ignorance of the genetic basis of the latter class is so great that there is no question of any attempt at gene therapy of these polygenic diseases in the foreseeable future. It should also be stated that it is totally unrealistic to consider possible attempts to manipulate more complex and even less understood characteristics which may have some genetic component such as behaviour or intelligence, even if anyone had any wish to attempt such an objectionable and unethical course.

Is gene therapy of well defined single gene diseases likely to be attempted in the future? The problems of gene insertion and expression are broadly similar to those mentioned previously for patient gene therapy and may well be overcome in a decade or so, but this does not necessarily mean it should be used. For many couples there will be a preferable alternative. Consider the situation where two parents are both heterozygous for the same serious genetic defect and therefore there is only a one in four chance that a particular embryo is homozygous and therefore affected. Presumably embryo gene therapy would not be attempted until it was established that the embryo was homozygous and at this stage most parents would probably prefer to terminate the pregnancy and conceive again rather than take the uncharted path of gene therapy.

In the long-term future however, technical problems of gene insertion and regulation may be solved by a combination of animal gene therapy experiments and increased knowledge of the human genetic material. Parents may then wish to attempt gene therapy rather than termination, particularly if the genetic defect is one which is not life-threatening but a condition which requires sustained medical attention and is likely to reduce the child's quality of life. The important point at the present seems to be that society should consider this future possibility and its ethical implications before it arises, rather than when it is technically feasible.

IVF of human eggs raises other possibilities which, although not involving gene therapy in a narrow sense, are sufficiently close to the subject to merit attention. One possibility is that an embryo could be cloned *in vitro* after fertilization to assess the possibility of it carrying a genetic defect. For example, a mother who had previously given birth to a child with Down's syndrome might choose to have her next embryo fertilized *in vitro*. The embryo could then be allowed to develop to the two or four cell stage, where one cell could be used to grow up a clone. This clone would be grown in culture until it could be used to assess by cytogenetic technique whether the embryo had a chromosome abnormality which is diagnostic of Down's syndrome. Meanwhile, another cell or cells of the embryo would have been stored deep frozen. If the embryo was not a potential Down's syndrome individual it could be unfrozen and transferred to the mother's uterus. A more controversial suggestion is that IVF cloning and freezing techniques could be used in the same way to select a child of particular sex on the basis of its chromosome complement. There are obvious ethical and sociological implications of this which need to be considered by society as a whole.

7 Genetic Manipulation of Plant Cells

7.1 Plant cell culture techniques

Techniques for growing plant cell cultures are now very widely used by botanists and plant biochemists. These techniques have been described by BUTCHER and INGRAM (1976) and will only be described very briefly here. In recent years there has been widespread interest in their use to manipulate the genetic material of plants and some developments in plant cell genetics will be considered in this chapter.

There are several rather different kinds of plant cell culture. The simplest type is a callus culture, which is growth in the form of an undifferentiated mass of several hundreds or thousands of cells. Callus cultures can be grown from explants of a wide range of plant tissues. They are grown on the surface of solid agar and may be subcultured by transferring fragments to fresh medium. Levels of plant growth hormones in the medium are adjusted to maximize the rate of cell division.

For genetical experiments, it is useful that plant cells can now also be grown as suspensions of largely individual cells. Such suspension cultures are usually initiated by transferring callus cultures into liquid medium and dispersing cells by agitation. Dispersion of cells is favoured by raising the auxin concentration or by adding low concentrations of cell wall degrading enzymes such as cellulase, but small aggregates of cells are usually also present along with the single cells (Fig. 7–1). Techniques have also been developed for producing colonies from single cells (or small aggregates) on the surface of the agar.

The use of plant cell cultures for genetical studies has a notable advantage over the use of animal cells. In many cases, whole plants can be regenerated from single cells growing in culture, i.e. the cells are said to be 'totipotent'. This opens up important opportunities for carrying out genetic manipulations on cells in culture and then growing them up into new types of plant for study or for agricultural use. This general statement should be qualified at present with the reservation that some agriculturally important crop plants (for example, several cereals) have proved rather difficult to regenerate from dispersed cell culture.

Another useful feature of plant cell culture which is not available in animal cell culture is that haploid cell cultures can be established from anthers. This is important because it gives the potential for isolating and studying recessive mutations without the complications of dominant alleles masking recessive mutant alleles. If cell division is blocked in haploid culture (by a brief exposure to the drug colchicine, for example) the chromosomes may fail to separate during mitosis and the cultures may become diploid. These will normally by homozygous for all genes and so regeneration of whole plants from such cell cultures can provide a much faster way of obtaining totally homozygous plants

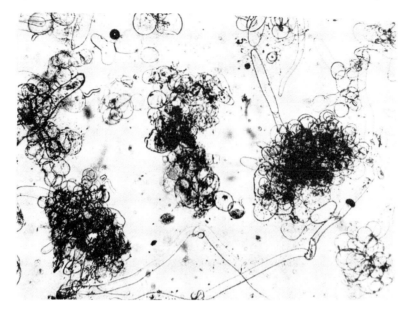

Fig. 7–1   A suspension culture of cells of *Catharanthus roseus*. (Courtesy of Professor W. W. Fowler, Wolfson Institute of Biotechnology, University of Sheffield.)

than repeated rounds of backcrossing which are necessary in conventional plant breeding.

7.2   Genetic variability in regenerated plants

If a cell culture is established from a plant and whole plants are regenerated from that cell culture, it is often noticed that there is a remarkably high frequency of variability of the phenotypes of plants which develop (Fig. 7–2). The phenomenon has been called 'somaclonal variation'. An example of this phenomenon is provided by one experiment where 72% of plants regenerated from rice callus cultures showed significant differences from controls in terms of plant height, morphology, chlorophyll or fertility. The genetic basis of this very high rate of variation in cell cultures is not entirely clear. It may be partly due to conventional mutation (i.e. changes in DNA sequences which constitute the gene) but the frequency is so high that other explanations are currently being considered. One possible explanation is that frequent, very small (often microscopically undetectable) rearrangements of chromosome regions may occur during the establishment of cell cultures and these may lead to changes in gene expression. This proposal requires further research before it can be accepted with confidence.

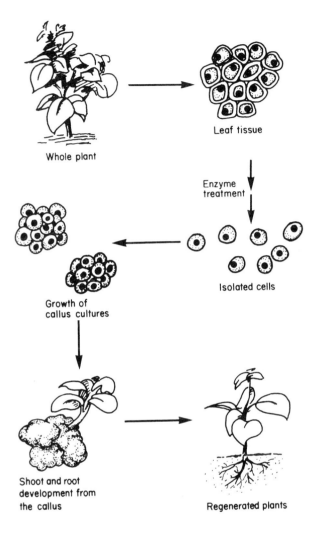

Leaf tissue

Whole plant

Enzyme
treatment

Isolated cells

Growth of
callus cultures

Shoot and root
development from
the callus

Regenerated plants

Fig. 7–2   The generation of somaclonal variation. Leaf tissue is isolated from a mature plant. Individual cells are obtained following enzyme treatment of the tissue and they are grown up to give many callus cultures. After some time in an appropriate culture medium the callus cultures develop a shoot, then a root, and may be planted in soil. These plants can then be examined for somaclonal variation.

Although the underlying mechanisms of somaclonal variation are not yet completely understood, it may be useful in the generation of new agricultural varieties. SHEPARD (1982) has studied the phenomenon in a variety of potato known as Russet Burbank, the most widely grown variety of potato in the U.S.A., accounting for 40% of total production. He has made the interesting observation that plants regenerated from callus culture of Russet Burbank vary considerably in their resistance to the fungus *Phytophthora infestans*, which caused the Irish potato famine in the 1840s and is still a major problem in commercial potato farming. Some regenerated plants have greater resistance to the fungus than standard Russet Burbank plants. In sugar cane which has been regenerated from cell culture, lines have been found with increased resistance to important diseases (such as downy mildew and eyespot disease) and also with increased sucrose content.

7.3   Direct selection of mutants in culture

Selection for certain classes of mutants may be made directly on plant cells, grown in culture, in the same way as animal cells (section 2.1), and on microorganisms. Because selection on cell cultures may be made on very large numbers of cells (tens of millions), there is a better opportunity for picking out rare variants than would exist if selection was made on hundreds or thousands of whole plants. If a cell culture mutant can then be used to regenerate a whole plant, a new variety may be produced with useful agricultural properties.

Selection of mutants in cell cultures has to be confined to phenotypic characteristics which are expressed in this situation. This limits the approach to characteristics which are expressed at the level of the cell rather than the whole organism. For example, if the aim is to obtain a drought tolerant variety of crop plant, it would obviously be impossible to select a mutant with longer roots using cell culture techniques, but it might be possible to select cells with altered osmotic control, which may produce regenerated plants with altered drought tolerance.

Many resistant mutants have been selected in plant cell cultures. R.S. Chaleff has selected tobacco cell resistant to the herbicide picloram by plating them in the presence of the herbicide. Regenerated plants had increased resistance conferred by dominant or semi-dominant genes. Increased herbicide resistance in crop plants permits more effective discrimination between the crop plant and weeds during treatment.

The nutritional quality of plant proteins may be limited by low levels of certain amino acids and therefore attention has been directed at possible ways of rectifying this. One approach to this has been to screen for cell culture mutants which are resistant to analogues of amino acids. These analogues are normally toxic because they are incorporated into proteins and render them non-functional. Plant cells may acquire resistance to an analogue by overproduction of the normal amino acid, which swamps the harmful effects of the analogue and reduces its incorporation into protein. Selection of mutants which are resistant to analogues therefore gives a way of obtaining mutants with high levels of particular amino acids. For example, cell culture mutants which are selected on the basis of resistance to the amino acid analogue hydroxylysine overproduced the normal amino acid lysine. Although this work is promising, it must be remembered that increasing the levels of deficient amino acids in the free amino acid pool is only the first step towards altering the amino acid content of the seed storage proteins which are used by humans as food. In addition to increasing the levels of free amino acids, it may be necessary to alter the genes coding for these storage proteins by other forms of genetic manipulation so that they have a higher proportion of code words for the required amino acids.

7.4   Protoplast fusion

An important feature of the genetics of animal cells in culture is that they can fuse to produce new combinations of genetic material (Chapter 2). Can plant cells fuse in an analogous way? The immediate answer to this question is that they are prevented from doing so by the elaborate cell wall which surrounds them, but E.C. Cocking and others have shown that protoplasts produced after removal of the cell wall *can* fuse and have potential for genetic studies.

Protoplasts can now be isolated from many species of plant and from many tissues (for example, leaves, roots, coleoptiles, pollen mother cells, callus cultures). The plant cell walls are removed by a variety of enzymes (cellulases, hemicellulases and pectinases), and are then maintained in growth medium in which the osmotic properties are very closely controlled. The frequency of protoplast fusions can be increased by various treatments, such as exposure to sodium nitrate or polyethylene glycol.

Protoplasts from different species can be fused together. If the fusing protoplasts are from closely related species, the chromosome complements of regenerated plants may have slight differences from the sum of the two original chromosome complements. For example, plants regenerated from fusions between *Petunia hybrida* protoplasts and *Petunia parodii* protoplasts had chromosome complements ranging from 24 to 28, although the sum of the two original complements is 28. However, if the fusing protoplasts are from less closely related species it is often observed that the chromosomes of one of them are eliminated during growth in callus culture. Plants regenerated from the callus may appear to contain chromosomes of only one of the species when examined microscopically. However, in some cases evidence has been found for the presence of small amounts of the genetic material of the other species, despite the absence of its chromosomes on visual examination. For example, D. Dudits fused protoplasts of carrot (*Daucus carota*) with *Aegopodium podagraria*. Regenerated plants had only carrot chromosomes but molecular biological techniques which detect particular types of nucleic acid sequences revealed the presence of genetic material from *Aegopodium*. Dudits speculates that after fusion the chromosomes from *Aegopodium* may have been broken into small fragments and incorporated into the *Daucus* chromosomes in a broadly similar way to the incorporation of chromosome fragments into mammalian chromosomes discussed in Chapter 3. This system requires further study, but it does suggest a possible way of inserting useful genes from one species into chromosomes of another.

7.5 Incorporation of new genes into plants

Is it possible to introduce foreign DNA into plant cells either directly or using some kind of a vector? There have been several reports of the uptake of free DNA into plant cells leading to changes in the plant's phenotype, but the experiments have been found to be irreproducible and the phenomenon remains poorly defined. In recent years, plant molecular biologists have concentrated their efforts on the use of vectors to carry foreign genes into plant cells.

The most promising vector is the Ti ('tumour inducing') plasmid (cf. section 5.1), a large plasmid carried by some strains of the bacterium *Agrobacterium tumifaciens*. When these bacteria infect plants, a mass of undifferentiated growth, called a crown gall tumour, develops at the junction of the root and the stem (the crown). A brief outline of the underlying molecular biology of the infection is necessary before considering the use of Ti plasmids as vectors (Fig. 7–3).

Bacterial infection causes the plant cells to start producing one or more unusual amino acids (not normally produced by plant cells) called opines. The commonest opines are called opaline and nopaline. Opines can be used as a sole nitrogen source for the bacterium and so the bacterium is effectively diverting much of the plant's metabolism for its own use. How is this brought about?

It can only be achieved if the infecting bacteria contain the Ti plasmid. Using molecular biological techniques, CHILTON (1983) and others have shown that part of the plasmid DNA becomes incorporated into the plant chromosomes during the initiation of a crown gall tumour. The bacterium is in effect carrying out its own form of genetic engineering on the plant by introducing new genes into the plant chromosomes which cause production of opines and proliferation of the opine producing cells.

This phenomenon clearly has potential for laboratory genetic engineering. Recombinant DNA techniques can be used to insert additional pieces of DNA into the plasmid DNA and they can then be taken into the plant chromosome with the fragment of Ti DNA during initiation of the crown gall tumour. Crown gall tumour cells are difficult to regenerate into whole plants but this has now been achieved with appropriate cytokinin levels. It is therefore possible to use the Ti plasmid to introduce foreign DNA into plants.

As with animal experiments (Chapter 5 and 6) it is essential to ensure that the inserted genes are expressed (i.e. transcribed into messenger RNA and then translated into protein). This does not normally occur with randomly inserted genes, but it has recently been shown in some experiments that if an existing gene is removed by biochemical means from the plasmid DNA and a foreign gene is inserted at *exactly* the same position, the new gene is expressed in crown gall tumour cells. This is because the adjacent DNA bases, which are essential for the regulation of genetic activity are still present and bring about regulation of the inserted gene in the same way as they would for the original one.

Some plant viruses are also being considered as possible vectors for introducing a fresh genetic material. The caulimoviruses (e.g. the cauliflower

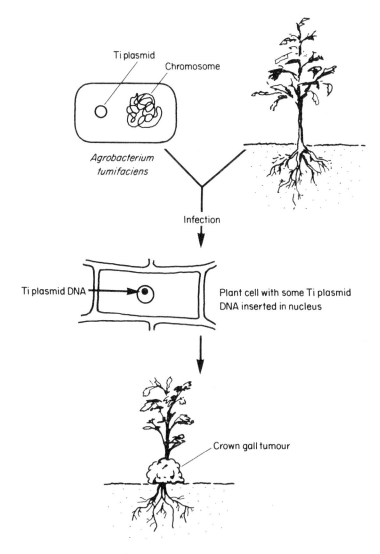

Fig. 7–3 Crown gall tumour formation following infection with *Agrobacterium tumifaciens*. Part of the DNA of the Ti plasmid becomes stably incorporated into the plant cell's chromosomes, leading to opine synthesis and cell proliferation.

mosaic virus) have been studied in particular detail because their genetic material is double stranded DNA (in contrast with many other plant viruses) and this makes them convenient for manipulation by recombinant DNA techniques. The proposed strategy for genetic engineering would be to infect whole plants with the virus carrying foreign DNA. Cauliviruses spread throughout the plant and, although the virus is not transmitted through seeds, it is transmitted during vegetative propagation.

The use of plant viral DNA as a vector presents a problem which is not encountered with the Ti plasmid. The viral DNA is packaged in a protein coat and this confers a limit to the amount of DNA which can be accommodated in virus particles. If new DNA is added by laboratory manipulation, some existing DNA must be deleted to stay within the size limit and deletion of some genes may harm viral functions. This problem, among others, indicates that although cauliviruses are still considered as potential gene vectors for plants, more basic research is needed to provide detailed information on their molecular biology.

One important point to note about both *Agrobacterium tumifaciens* and cauliviruses is that neither of them infect monocotyledenous plants and therefore they do not seem suitable for genetic engineering with many important crop plants. Research into ways of extending the range of infection of possible vectors is therefore of particular importance for the future of plant genetic engineering.

If suitable vectors are developed, what types of genes may be inserted into crop plants? Many important traits in crop plants are not specified by single Mendelian genes, but are produced by interaction of tens or hundreds of genes. For example, adaptations to drought or high temperatures (such as a reduction in leaf surface area or fewer stomatal openings) have a complex genetic basis which is not understood. Even such an attrative proposal as the introduction of nitrogen fixing genes into crop plants presents formidable difficulties. There are seventeen genes for nitrogen fixation in *Klebsiella pneumoniae* which are under an elaborate system of control. Insertion of prokaryotic genes into eukaryotic cells will lead to defective control of gene activity. Other features of plant cell metabolism may have to be modified for effective nitrogen fixation; the enzyme nitrogenase which plays a central role in nitrogen fixation is highly oxygen-labile and therefore the level of oxygen within the cell must be closely regulated.

The prospects for introducing simpler characters into plants are brighter. Herbicide resistance is often a single gene character and is a likely candidate for future genetic engineering experiments. Insertion of genes coding for modified seed storage proteins with improved amino acid composition is also a real possibility, although problems will arise here because seed storage proteins are a combination of slightly different proteins coded for by a family of related genes in each plant. Therefore several genes may have to be replaced before a significant impact is made on the overall nutritional quality of the seed proteins.

The aim of this section has been to present a general view of current work in this area and the overall picture may seem rather pessimistic. However, current problems should be put in a wider context. Effective research has only been carried out with possible plant gene vectors for about ten years. Although major problems face plant genetic engineers, most of them are likely to be overcome in the next few decades. When considered in that time scale, plant genetic engineering seems sure to contribute significantly to increasing world food production.

Further Reading

ANDERSON, W.F. and DIACUMAKOS, E.G. (1981). Genetic engineering in mammalian cells. *Scientific American*, **245** (pt. 1), 60–93.

BERG, P. (1981). Dissections and reconstructions of genes and chromosomes. *Science*, **213**, 296–302. (A discussion of new vectors for cloning genes in mammalian cells.)

BUTCHER, D.N. and INGRAM, D.S. (1976). *Plant Tissue Culture*. Studies in Biology, No. 65. Edward Arnold, London.

CHILTON, M. (1983). A vector for introducing new genes into plants. *Scientific American*, **248** (pt. 6), 36–45.

DAY, M.J. (1982). *Plasmids*. Studies in Biology, no. 142. Edward Arnold, London.

KLOBUTCHER, C.A. and RUDDLE, F.H. (1981). Chromosome mediated gene transfer. *Annual Review of Biochemistry*, **50**, 533–55.

LEWIN, B.L. (1980). *Gene Expression. Volume 2, Eucaryotic Chromosomes*, 2nd edition. John Wiley and Sons, New York. (Chapters 6, 8 and 9 discuss the genetics of mammalian cell cultures.)

MOTULSKY, A.G. (1983). Impact of genetic manipulation on society and medicine. *Science*, **219**, 135–40.

OLD, R.W. and PRIMROSE, S.B. (1981). *Principles of Gene Manipulation. An Introduction to Genetic Engineering*, 2nd edition. Blackwell Scientific Publications, Oxford. (Chapters 9 and 10 discuss cloning in higher organisms. Other chapters give background information on gene cloning in bacteria.)

RIGBY, P.W.J. (1982). Expression of cloned genes in eukaryotic cells using vector systems derived from viral replicons. In *Genetic Engineering*, **3**. (WILLIAMSON, R., ed.). Academic Press, London and New York.

SHARP, J.A. (1977). *An Introduction to Animal Tissue Culture*. Studies in Biology, No. 82. Edward Arnold, London.

SHEPARD, J.F. (1982). The regeneration of potato plants from leaf-cell protoplasts. *Scientific American*, **246** (pt. 5), 112–21.

SHEPARD, J.F., BIDNEY, D., BARSBY, T and KEMBLE, R. (1983). Genetic transfer in plants through interspecific protoplast fusion. *Science*, **219**, 683–8.

WILLIAMSON, B. (1982). Gene therapy. *Nature*, **298**, 416–18. (A thoughtful discussion of the medical implications of gene therapy.)

YELTON, D.E. and SCHARFF, M.D. (1981). Monoclonal antibodies: a powerful new tool in biology and medicine. *Annal Review of Biochemistry*, **50**, 657–80.

Index